Teaching Your Kids New Math, K-5

for
dummies®
A Wiley Brand

Teaching Your Kids New Math, K–5

by Kris Jamsa, PhD[2]

A Wiley Brand

Teaching Your Kids New Math, K–5 For Dummies®

Published by: **John Wiley & Sons, Inc.,** 111 River Street, Hoboken, NJ 07030-5774, www.wiley.com

Copyright © 2022 by John Wiley & Sons, Inc., Hoboken, New Jersey

Published simultaneously in Canada

For general information on our other products and services, please contact our Customer Care Department within the U.S. at 877-762-2974, outside the U.S. at 317-572-3993, or fax 317-572-4002. For technical support, please visit https://hub.wiley.com/community/support/dummies.

Wiley publishes in a variety of print and electronic formats and by print-on-demand. Some material included with standard print versions of this book may not be included in e-books or in print-on-demand. If this book refers to media such as a CD or DVD that is not included in the version you purchased, you may download this material at http://booksupport.wiley.com. For more information about Wiley products, visit www.wiley.com.

Library of Congress Control Number: 2022934351

ISBN: 978-1-119-86709-8 (pbk); 978-1-119-86710-4 (ebk); 978-1-119-86711-1 (ebk)

SKY10033827_032922

Contents at a Glance

Table of Contents

Introduction

People use math every day: to shop, to decide when they need to stop for gas, to know when to tune in to watch their favorite team play their division rivals, to pinch pennies, and more.

Math is not new. Newton used math to understand gravity. Columbus used math to find a new world, and Einstein used math to explain the workings of the universe. Don't worry; all those topics are covered in different books. This one is focused strictly on math.

With all the great things that math has given us throughout history, you may be asking, "Why do we need *new* math?"

It turns out that math, like a fine wine, improves over time. New math, therefore, is a result of finding better ways to solve problems.

The good news is that we call it "new math," not "hard math." New math techniques for solving addition, subtraction, multiplication, and division problems are easier for kids (and parents) to learn and master.

This book provides step-by-step instructions for how to use both old and new math to solve problems that kindergartners through fifth-graders must know. It also provides instructions, examples, and practice problems, and sometimes suggests what you should say as you teach your child.

About This Book

There are many reasons why you may have chosen this book. Perhaps your child just asked you for help with their math homework and you experienced a new math "uh-oh" moment. Maybe you just left a parent-teacher conference with practice problems you don't understand, or perhaps, your little math whiz is ready for math at the next grade level.

Regardless of your reasons, you've got the right book.

This book presents the math your child must know from kindergarten through fifth grade, with each chapter focusing on specific key concepts. If your child needs help with only fractions or telling time, you can turn to a specific chapter that addresses that topic. If your child is struggling at their current grade level, you can take a step back and strengthen their foundation, knowledge, and confidence from previous grades.

Within each chapter, you will find step-by-step instructions for how to teach each concept. I've also provided many example problems for you to work through with your child. Let them solve the problems right on the book's pages if you want — it's your book after all.

Foolish Assumptions

I like math — math works! I usually get correct change at the grocery store, I can identify the largest slice of pizza, and I know when the bathroom scale must be broken. That said, I recognize that math may not be everyone's favorite thing to do!

Relax. If it's been a while since you've done math without the help of your phone's calculator app, you'll be fine. In fact, you may surprise yourself with what you remember!

Don't let the phrase "new math" intimidate you. This book covers good ol' addition, subtraction, multiplication, and division, just in new ways. You *can* teach an old dog new tricks, and you *can* quickly learn new ways to solve old problems!

Icons Used in This Book

TIP

The Tip icon marks tips (duh!) and shortcuts that you can use to make learning new math easier, and sometimes when to know that it's time to take a break!

REMEMBER

The Remember icon points out things that you should, uh, remember! You and your child will examine a lot of topics throughout this book. I include this icon for those things you should keep in the back of your mind as you move forward.

Beyond the Book

Becoming a "math whiz" requires practice. I've sprinkled many math problems your child can complete throughout this book. But because I know "practice makes perfect," I've provided many worksheets of problems on this book's companion website at `www.dummies.com/go/teachingyourkidsnewmathfd`. As you and your child work your way through this book, you should take the time to download and print the corresponding practice worksheets.

In addition, should you need quick help on a math process and you don't have this book handy, I've created a cheat sheet you can download and print that will help you with many key concepts. You'll also find it at `www.dummies.com/go/teachingyourkidsnewmathfd`.

Where to Go from Here

This book's chapters present kindergarten through fifth-grade math in order. However, many math problems occur because kids don't have a strong foundation in concepts taught prior to their current grade levels. I recommend that you review the math from earlier grade levels with your child. In so doing, you will strengthen their knowledge and build their confidence in what they know! In fact, it may even make math fun!

Lastly, if you are pressed for time and your child just placed the homework due today next to your coffee and orange juice, you can jump directly to the corresponding topic in this book to save the day. Later, when you have a few more minutes, you and your child can review the rest of the chapter.

1

Laying Down the Basics with Kindergarten Math

IN THIS CHAPTER. . .

» Building your teaching confidence

» Taking a deep breath and getting started

» Justifying your investment of time

» Establishing a teaching plan

» Planning your teaching road map

Chapter **1**

Unleashing Your Inner Math Teacher

People buy books to help their kids with homework for many reasons. Some want to give their child an advantage, some are tired of the daily homework struggles, and many, who didn't even want to remember "old math," find "new math" unrecognizable. Whatever your reason for buying this book, relax! If you fall into this last category, then this book is your guide. In it, I sometimes tell you what to say, not just to teach your child "new math," but also to build their confidence along the way!

People often say they don't like math because they have never really understood it, but you don't need to be a math scholar to teach your child kindergarten through fifth-grade math. Perhaps Albert Einstein said it best:

Do not worry too much about your difficulties in mathematics. I can assure you that mine are still greater.

I'm Not Sure I Remember "Old Math," and Now There Is "New Math"

If you are like most people, you use the calculator app on your phone to add, subtract, multiply, and divide, and it's been a while since you had to perform long division or reduce fractions. Don't worry. In this book, I walk you through each process so you can present and teach it to your child with confidence. You'll probably amaze your friends and family with your newfound expertise, and you may even surprise yourself!

If you were born before 2010, you learned to add numbers by carrying, subtract numbers by borrowing, and multiply numbers without the use of boxes. Here's an example of old-school multiplication:

$$
\begin{array}{r}
^{2} \\
37 \\
\times\,23 \\
\hline
111 \\
\underline{740} \\
851
\end{array}
$$

You've likely used these techniques successfully throughout your adult life — that is, until your child asked you for help with their homework! Carrying and borrowing aren't how your child adds and subtracts. Instead, they use approaches that may seem more confusing than a foreign language. Here's the same multiplication problem solved with "new math" boxes:

$$
\begin{array}{r}
37 \\
\times\,23 \\
\hline
\end{array}
$$

	30	7
20	600	140
3	90	21

$$
\begin{array}{r}
^{1} \\
600 \\
+140 \\
+\;90 \\
\underline{+\;21} \\
851
\end{array}
$$

Take a deep breath! I like old-school math, and it has worked well for me. But, unlike an old dog, I am willing to learn new tricks, and I must admit that the "new math" techniques are easy, accurate, and fast! In this book, I present you with how to teach your child not only the old-school techniques, but also how to master the new math as well.

TIP

Traditions die slowly. That said, you need to be open to the new ways of learning — especially if you want to help your child master math. My goal with the techniques in this book is to make it enjoyable for your child to learn new math with you. The bond you will create with your child is possibly more important than establishing their math foundation for future success.

Old Math, New Math, Common Core Math

Years ago, smart people in each state would get together and establish the state's learning curriculum — the things teachers in that state would teach. The problem was that each state's curriculum was different. What a first-grader learned in California might be different from what a first-grader learned in Alaska or Wyoming. Simultaneously, math scores within the United States were falling. In fact, as of 2015, math scores in the United States had fallen from first to thirty-fifth in the world!

In 2009, the National Governors Association and the Council of Chief State School Officers got together to create the Common Core State Standards Association. From that group, Common Core Math was born.

If you ask a roomful of educators to comment on Common Core, you will hear a wide range of opinions. Some love it! They want to see standards across grade levels and across the country. Others hate it. They want the government to leave curriculum decisions to the individual states. This book does not debate for either side. Instead, I simply present the math skills these groups identified as important for your child to know and for teachers to teach.

It's Not Too Late to Get Started

It's great that you've picked this book. If your child is struggling with math or you are struggling to help them, this book can help solve the problem. If your goal is to help your child get ahead, you can help them master the skills for their current grade level and then move on.

If your child is struggling today, it's likely because they didn't master the skills in a previous grade level. That problem is easy to solve. This book starts with kindergarten math. You may want to start there regardless of your child's current grade level. Depending on your child's age and current skills, you may move through that content quickly. The successes your child will experience will give them greater confidence in their knowledge, and you may find that you're able to fill in a few key gaps.

In any case, if your child is having trouble at their current grade level, you can simply turn back a few pages to a previous grade level and lay a better foundation. Remember, you paid for the entire book. Use it!

You may be worried that you are too busy to help your child with math or that you can't learn new math. Relax. You have the right book. Raising kids can be hard. The good news is that teaching math is not. You can do this!

Your Return on Investment

Kids who do well in math tend to do well in school. Conversely, kids who struggle in math often struggle in school. Math is important. That's why schools teach math every day and in every grade.

Students who do well in school tend to go to college, and college graduates tend to earn more than $1 million more in their lifetimes than workers without a college degree. The time investment to work through this book is about 15 minutes per day. With the potential for an extra $1 million at stake, that's a great return on the investment of your time!

Establish Your Routine

A key to your success in teaching your child math is to establish a daily routine that works for both of you. You might, for example, practice flash cards before breakfast and work on other math problems before dinner.

By establishing a routine, you will find that you can make time and that your child has the expectation that you will be working together. Knowing that you care about their success is important to your child.

Keep a Positive Attitude

Mathematical problems have a right and wrong answer, and your child will sometimes make math errors. The key is how you respond in such situations. Be positive about the problems your child gets right, and remain positive about what they can learn from the problems they get wrong. You will find that a positive attitude about math is contagious, and being positive will help your child succeed.

REMEMBER

Have fun! You are setting out on an adventure that will forever change your child's life.

Chapter **2**

Knowing the Number Names 1 to 9

Numbers and counting are important. Without them, musicians could not keep the beat, sports teams could not keep score, and Neil Armstrong could not have taken "One small step for man." Without numbers, there'd be no speed limits — wait, that might be kind of cool, not that you want to tell your kids that. But anyway, counting requires numbers, and in this chapter, your child is going to start learning them. So, let's get going; it will be as easy as 1, 2, 3!

TIP

Before you get started, you'll want to have the following supplies on hand:

>> Box of 100 straws

>> Deck of 3x5 index cards

>> Deck of playing cards

Preparing for a Lesson

Before you start teaching your child, first prepare a few key teaching tools. To start, write the numbers 1 through 9 on individual 3x5 index cards, as shown below:

| 1 | 2 | 3 | 4 | 5 | 6 | 7 | 8 | 9 |

If you don't have 3x5 index cards, sheets of paper will do — however, because you'll be practicing daily, the index cards may be more durable and easier to store. Next, I recommend using straws as objects your child can count. They're also sturdy, and colorful straws can make the activity more fun. You can buy boxes of 100 (or more) straws from the grocery story, a wholesale store, or an online retailer such as Amazon.com. If you don't have straws, you can substitute pennies or any other small objects. Later, I suggest that you use playing cards, so if you don't already have some, you may want to pick some up while you are shopping.

TIP

Try to pick a quiet location where you and your child can work without distraction such as the kitchen table away from a TV.

1. **Place the index card with the number 1 face up on the table and give your child one straw.**

Tell your child, "This is the number 1, and you have one straw."

2. **Repeat this process with the index card with the number 2, and place the second straw in front of your child on the table.**

Place the card with the number 1 next to the card with the number 2:

| 1 | 2 |

3. **Ask your child to point to the number that matches the number of straws they have.**

They should point to the number 2. If not, take one straw away from them and say, "You pointed to the number 1. You had two straws, but now you have one."

4. **Place the card with the number 3 on the table and pick up three straws:**

| 1 | 2 | 3 |

5. **Count the straws out loud, "one, two, three," as you place them in front of your child.**

Point to each card, again counting out loud.

6. Ask your child to point to the number 3.

7. Repeat this process for the numbers 4 and then 5:

1	2	3	4	5

8. Ask your child to point to the number 1 and ask them to pick up one straw.

As they pick up the straw, have them count out loud. Repeat this process for two straws, then three, four, and five. Then ask your child to put down the straws.

9. Point to any number and ask your child to pick up the corresponding number of straws, counting the straws out loud as they take them.

10. With the five index cards facing up, present one to five straws to your child, allowing them to count them out loud.

Then have your child point to the corresponding number on the index cards.

11. Repeat this process until your child masters the numbers 1 through 5 — which may take a few lessons.

As your child starts with counting, count the numbers out loud with your child as they count.

TIP

Learning math takes time and patience. Pay attention to when your child has had enough, and always try to end the lesson on a successful note. As a general rule, 10-minute sessions work well for most concepts.

TIP

In your next lesson, you can add the numbers 6 and 7:

1	2	3	4	5	6	7

After your child masters the numbers 1 through 7, you can repeat this process for the numbers 8 and 9.

1	2	3	4	5	6	7	8	9

Practicing Counting with Flash Cards

One of the best ways to master essential math skills is to practice them regularly. Flash cards are a great learning tool.

Using the numbered index cards, place the cards in order from 1 through 9.

1. **Start with the number 1 and present the card to your child, having them say the number out loud.**

   Repeat the process one or two times as you present the cards in order.

2. **After your child has mastered the ordered numbers, shuffle the cards and present the cards one at a time out of order.**

TIP

If you do not have index cards, you can point to the numbers printed here, first in order and then out of order:

1	2	3	4	5	6	7	8	9

Also, if you happen to have the Uno card game, you can use the cards to practice counting with your child.

Counting different objects

Throughout the day, there are many opportunities for your child to master counting through the numbers you have practiced thus far. You might, for example, count your child's books, the number of socks in the laundry, the number of plates on the table, and so on. Find opportunities for your child to count. The more you practice with your child, the stronger their counting skills will become.

Pointing out numbers

TIP

As you encounter numbers throughout your day, point out the numbers to your child and have them say the numbers out loud. You might count the numbers on your phone, prices at the grocery store, gas prices at the pump, or even the page numbers of this book. When you encounter large numbers, such as a price ($19.95), have your child say the numbers one at a time: "One, nine, nine, five."

Writing the Numbers 1 through 9

After your child can read and recognize the numbers 1 through 9, they can learn to write each of them. Worksheet 2-1 at this book's companion website (www.dummies.com/go/teachingyourkidsnewmathfd) is good practice for tracing and printing the numbers 1 through 9, as shown in Figure 2-1.

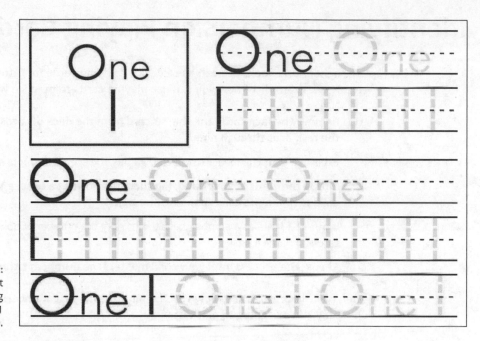

FIGURE 2-1:
Use a worksheet
to practice writing
the numbers 1
through 9.

Allow your child to write the numbers within this book. Also, as your child traces the numbers on worksheets you print, your focus should be on whether they draw the correct number rather than on using perfect handwriting.

TIP

Counting Objects and Writing the Corresponding Number

After your child knows the numbers 1 through 9, they should demonstrate that they can transfer their understanding by counting objects on a worksheet and writing the corresponding number.

Download and print Worksheet 2-2 from this book's companion website and help your child count the pictures out loud and write the corresponding number.

Identifying Numbers on Playing Cards

Throughout this book, you will use a deck of playing cards for different purposes. Here's one example of how you can use playing cards to practice with your child:

1. **Remove the face cards and the '10' card from the deck of cards, leaving the rest, aces through nine.**

   Explain to your child that in most card games, the ace card counts as 1.

2. **Mix up the cards in your hand, flip them over one at a time, and ask your child to name the corresponding number.**

   If your child misses a number, have them count the number of objects on the card and name it again.

3. **Repeat the process until your child masters the numbers 1 through 9.**

Then place your cards face down on the table (hiding the numbers), and tell your child that you are going to play a game like so:

1. **Tell your child that they will turn over a card.**

2. **If your child can name the number correctly, they get to keep the card. Otherwise, you get to keep the card.**

   Tell your child that they can count the objects on the card to help.

3. **After your child names the last card, the one with the most cards wins!**

TIP

If your child has difficulty with all the numbers in the card deck, work first with the cards for the numbers 1 through 5. After your child has mastered those numbers, you can move on to additional numbers, adding first the cards for 6, then 7, and so on.

Counting Numbers on a Number Line

Later in this book, you will use number lines to teach your child to add and subtract. When your child knows the numbers by sight, it's a great time to introduce a number line, as shown here:

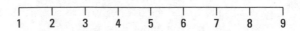

Introduce the number line to your child by telling them that a number line shows the numbers 1 through 9 in numerical order and that they can use it to help them count.

TIP

As your child works with the number line, have them touch the corresponding numbers on the line as they count.

1. **Have your child count the numbers 1 through 9 using the number line.**

2. **Point to a number and ask your child to start counting from that number through 9.**

   For example, point to the number 4 and ask your child to start counting from there.

3. **After your child can successfully count to 9 from different numbers, say to your child, "Let's try counting without using the number line."**

   Start by saying "one" out loud with your child. Then say, "one, two."

4. **Repeat this process until your child can count from 1 through 9.**

   If your child can't remember a number, allow them to look on the number line.

The next step is to present number lines with missing numbers, like the one I show here, and to explain that the missing numbers need to be filled in. Help your child with the first few and then allow them to complete the rest.

TIP

If your child has difficulty with a number, have him or her count from 1 up to the missing number.

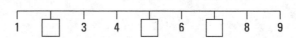

Learning Your Phone Number

For safety reasons, it's very good for your child to know your phone number. Also, your child will find it great fun to know that they know your number!

TIP

Many phone numbers have the number zero, a concept covered in Chapter 7. For now, simply introduce the number to your child by name: "This number is zero; you will learn more about it soon."

1. **Write your phone number (including the area code) on a piece of paper.**

2. **Have your child start at the first number and say the digits of your phone number as they are written.**

3. **Say the first three numbers out loud and have your child repeat them without looking at the paper.**

   Repeat this process several times.

4. **When your child has mastered the three digits of the area code, focus on the next three numbers by saying and repeating them.**

   Have your child say the first three numbers from memory and then read the three numbers that follow from the paper. Repeat this process a few times and then say the numbers from memory with your child.

5. **Repeat this process with the last four digits.**

TIP

   If you have a second phone available, allow your child to type in the numbers and call you!

Allow your child to keep the paper with your phone number. Repeat this process for several lessons until your child has memorized your number.

Chapter **3**

Moving on up to 20

n Chapter 2, you teach your child to recognize and count numbers 1 through 9. In this chapter, you're going to use a similar method to teach your kids the same process but to count to 20 — a key kindergarten curriculum standard. With all the numbers in the world, 20 feels a little random. Why stop at 20? My guess has something to do with 10 fingers and toes. So, take off your shoes and socks and get ready to count to 20.

TIP

Before you get started, you'll want to have the following supplies on hand:

>> Box of 100 straws

>> Deck of 3x5 index cards

>> Deck of playing cards

Tiptoeing into the Teens

Your approach to teaching the numbers 10 through 20 will be very similar to teaching the numbers 1 through 9. Before you start this lesson with your budding math genius, write the numbers 10 through 20 on your 3x5 index cards (or sheets of paper). You will also need 20 straws (or other objects) for counting.

Follow these steps to begin working on the numbers 10 through 20:

1. **Review the numbers 1 through 9 by laying out the cards in order as shown here.**

1	2	3	4	5	6	7	8	9

2. **Point to each of the cards and have your child say the corresponding number out loud.**

3. **Place the 10 card next to the 9 card and count out 10 straws.**

   Tell your child that the number 10 follows 9. Ask your child to count the straws, helping them say "ten!"

4. **Repeat this process, introducing 11 and asking your child to say "eleven."**

   Have your child count the 11 straws. Then point at each card, starting with 1, and have your child say the numbers out loud.

5. **Repeat this process for the numbers 12 through 15.**

   Count out loud from 1 through 15 with your child. If they have trouble with a number, have them refer to the cards, which should still be on the table. You should do the numbers 16 through 20 in another session.

When your child seems to have a grasp on reciting the numbers 1 through 15 in order using the flash cards as a prompt, you can move on to practicing the numbers from memory.

Place the cards for numbers 1 through 15 on the table in order, face up, and follow these steps:

1. **Have your child point to each card and say each number.**

2. **Turn over the 1 card so it is face down and repeat the process of counting all the numbers 1 through 15, having your child say 1 from memory.**

3. **Turn the 2 card face down and start the counting process again.**

   This time, your child is recalling both 1 and 2 from memory because both cards are face down.

4. **Repeat Steps 2 and 3 by turning over each consecutive number card and having your child recite all the numbers starting at 1.**

   If you child hits a wall and is having trouble remembering the numbers, you should stop for the day, celebrating the numbers they know from memory.

Working with a number line for numbers 1 through 15

When you teach a new concept, it's often helpful to review what your child already knows to build their confidence. That's why this lesson begins with a short review of the numbers 1 through 9. Then, you'll introduce the numbers 10 through 15. Don't worry, you'll get to 20 — you're just going to do it in steps.

Following is a number line with the numbers 1 through 15. Introduce the number line to your child and state each number out loud as you point to it.

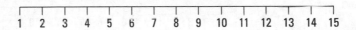

Spread the flash cards for numbers 1 through 15 randomly across the table; then follow these steps:

1. **Tell your child that they are going to place the cards back in order.**

   Tell them that they can use the number line to help.

2. **Ask your child to tell you which number comes first.**

   Then ask your child to look at the number line and find the card that comes next in order.

3. **Repeat this process for the remaining numbers.**

After your child has mastered numbers with the preceding process, mix up the cards again. This time, you're asking your child if they can put the numbers in order without using the number line by using the following steps:

1. **Ask your child, "What number do we start with?"**

   Have your child find the 1 card among the cards on the table.

2. **Have your child continue counting and placing the cards in order.**

3. **If your child gets stuck, tell them the number that comes next.**

   For example, you can say, "The number 6 comes after 5. Which card has the number 6?"

4. **Ask your child to count from 1 to 15 from memory.**

   Help your child as necessary.

Identifying matching numbers on a number line

I've provided some number lines from 1 to 15 that are missing several numbers. Help your child complete the first number line and then ask them to complete the rest by filling in the missing numbers.

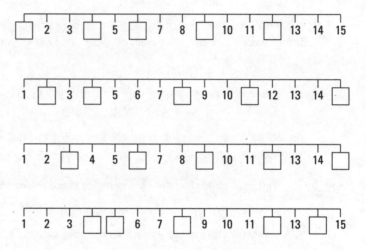

Reviewing Your Phone Number

In Chapter 2, you taught your child your phone number. Ask them if they still remember your phone number. If so, you might let your child call you. If they do not remember your phone number, then write it on a piece of paper, first reviewing the first three numbers, then the second three, and finally the last four.

Going All the Way to 20

Your process for introducing numbers 16 through 20 to your child will be the same method you used to work through numbers 10 through 15. With your 3x5 index cards and straws in hand, you're ready to work through the following steps:

1. Place the cards with the numbers 1 through 15 in order on the table.

2. Show your child the card with 16 and say, "This is 16. It comes after 15."

Place the 16 card to the right of the 15 card and then count out 16 straws, placing them in front of your child. Ask them to count the straws out loud.

3. **Repeat this process for the numbers 17 through 20.**

4. **Say the numbers from 1 through 20 out loud with your child, pointing to a card with a number when necessary.**

Place the flash cards for numbers 1 through 20 on the table in order, face up, and follow these steps:

1. **Have your child point to and say the number on each card.**

2. **Turn the 1 card face down and repeat the process of counting all the numbers 1 through 20, having your child say 1 from memory.**

3. **Turn the 2 card face down and start the counting process again.**

4. **Repeat Steps 2 and 3 by turning over each consecutive number card and having your child recite all the numbers starting at 1.**

   If your child gets stuck, tell them "great job" for the numbers they have learned today and pick up the process from step 1 the next time you practice.

Counting the Numbers 1 through 20 Using a Number Line

Here is a number line with the numbers 1 through 20:

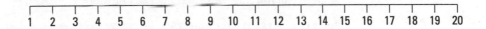

Present the number line to your child. Say each number out loud as you point to it.

Worksheet 3-1 at www.dummies.com/go/teachingyourkidsnewmathfd includes number lines from 1 through 20 with several missing numbers. Help your child resolve the first few missing numbers and then ask them to complete the rest. They can use the number line as needed.

Reviewing the Numbers 1 through 20 Using Flash Cards

Place your 3x5 index cards on the table, in order, from 1 through 20. Present each card to your child and ask them to say the number out loud. After your child has mastered the numbers in order, mix up the cards and have your child arrange them in numerical order.

TIP

Flash cards work when you use them repeatedly. Try to find time each day to review the numbers 1 through 20 with your child. They will quickly master the numbers and build confidence in the process.

Counting the numbers from 1 through 20 forward and backward

So far, you've taught your kiddo to count forward. Easy enough, right? Now it's time to introduce the idea of counting backward. This concept might be a little harder for kids to understand. Unless you're thinking of becoming an astronaut, you may wonder why one needs to count down numbers. The answer is subtraction, which is coming soon!

Place the 3x5 cards from 1 through 20 in front of your child. You can make two rows if necessary:

1	2	3	4	5	6	7	8	9	10
11	12	13	14	15	16	17	18	19	20

Then follow these steps:

1. **Have your child count the numbers out loud from 1 through 20.**

2. **Point to the number 20 as you say the number.**

3. **Point to and say the number 20 followed by 19: "Twenty, nineteen."**

4. **Repeat this process to count backward through all the numbers.**

Have your child practice counting the numbers forward and backward while looking at the flash cards. Then, repeat this process using the number line shown earlier in this chapter. Then, turn over the card with the number 20 and have your child say 20 followed by the numbers 19 down to 1. Next, turn over the card with the number 19 and have your child say 20 and 19 from memory followed by the remaining cards. Repeat this process of turning over one card at a time and saying the numbers until all the cards are face down.

TIP

Repeat this process for a few lessons. Your goal is for your child to count from 1 through 20 by memory and then also to count backward from 20 through 1.

Identifying and writing missing numbers from 1 through 20

Worksheet 3-2 at Dummies.com includes rows of numbers from 1 through 20 that contain missing numbers. Help your child solve the first few missing numbers and then ask them to complete the rest. Allow your child to use the number line shown earlier in this chapter as necessary.

Counting objects from 1 through 20 using a worksheet

Using straws and index cards to count gets boring after a while. Give your child a break and have them count other items. Worksheet 3-3 is a series of illustrations of various numbers of objects that your child can count, as well as a box in which they can then write the corresponding number. Help your child with the first few and then ask them to complete the rest.

Learning the Relationship between Numbers 1 through 10 and 11 through 20

After your child knows their numbers through 20, you can introduce the idea that numbers repeat, in a sense. The key to learning to count easily beyond 20 is for your child to know the relationship between numbers such as 11, 21, 31, 41, and so on. In this section, you lay that foundation.

Using your 3x5 index cards, place the cards from 1 through 10 in one row and the numbers 11 through 20 beneath them, as previously shown. Then follow these steps:

1. **Point to the numbers 1 and 11 and show your child that both numbers have a 1.**

2. **Repeat this process for the numbers 2 and 12, 3 and 13, and so on.**

3. **Pick up the cards with the numbers 14 and 18, hand them to your child, and ask them to put the cards back in the correct locations.**

4. **Repeat this process for the numbers 15 and 19, 13 and 17, and 11 and 1.**

5. **Spread the cards randomly about the table. Then, pick up any card from 1 through 9 and ask your child to find the card that has the same digit.**

Counting by Twos through 20

Back in the day, after Little League baseball games, each team would chant, "2, 4, 6, 8, who do we appreciate?" and cheer the other team's name. Setting aside the cheer, being able to count by twos will develop your child's understanding of numbers and patterns. In this section, you help your child learn to count to 20 by twos.

Place your 3x5 cards with the numbers 1 through 20 in front of your child. Before beginning the following steps, say something like the following to your child: "You know how to count from 1 to 20 by ones. Now you're going to learn to count to 20 by twos!" At school, they often call this process "skip counting."

1. **Point to the card with the number 2 and say, "Two."**

2. **Point to the cards with the numbers 2 and 4, and say, "Two, four."**

3. **Repeat this process and include the 6 card.**

4. **Ask your child what number they think comes next.**

    Repeat this process for the numbers 8 through 20.

5. **Try reciting the numbers along with your child.**

    If your child seems stuck and pauses, point to the correct number card and continue.

Your goal is for your child to count from 2 through 20 by memory. You will practice this task for several sessions.

Where To? Knowing Your Address

For safety reasons, it's important for every child to know their address. When you introduce the idea of knowing your address, tell your child that if they ever get separated from you, they should tell a police officer their address so the officer can bring them home.

1. **Write down your street address.**

    Don't worry yet about the city and state; you can review those another time.

2. **Tell your child that every house or apartment has a unique address so that people can find it and mail and packages can be delivered to the right place.**

3. Show your child your address and say it out loud to them: "We live at. . ."

4. Ask your child to memorize the numbers and then add the street name (and apartment number if necessary).

TIP

Practice your address with your child over several sessions. For fun, you may want to pull up Google Earth within your web browser and help your child to look up your address. Also, you might have your child write a letter and mail it to themselves using their address.

Chapter **4**

Comparing Numbers: Understanding More and Less

Adapting to change can be hard, which is why you may be struggling with helping your child with new math. The good news is that you can still count on numbers. They're still the ones you grew up with, and 2 is still greater than 1. In this chapter, you find out how to teach your child how to compare numbers. Bigger may not always be better, but at least your child will know which number is the bigger one!

TIP

Before you get started, you'll want to have the following supplies on hand:

» Box of 100 straws

» Flash cards you previously made for numbers 1 through 20

» Deck of 3x5 index cards

» Deck of playing cards

Understanding More, Less, and Equal

Before you teach your child how to compare numbers, you should teach them the concepts of more, less, and equal. It helps to give your child a visual example. Try the following steps:

1. Take two glasses of the same size and fill one glass about $\frac{2}{3}$ full and the other $\frac{1}{3}$ full.

   Set the glasses side by side and ask your child which glass has more water.

2. Ask your child to point to the glass that has more water first and then to point to the glass with less water.

3. Explain to your child that when one glass has more water than another, we say that the glass has the "greater" amount of water.

4. Ask your child to point to the glass with the greater amount of water and then ask your child to point to the glass with less water.

5. Add water to the less-full glass until the glasses have the same amount, and again set the glasses side by side.

6. Explain to your child that when the glasses have the same amount, we say the glasses are "equal."

Bringing out the straws!

Comparing numbers will be a new concept to your child, but comparing the quantity of things is probably something they're used to. Here's a way to introduce comparisons using your straws (or other objects):

1. Place two piles of straws on the table.

2. Ask your child to count the straws and say which pile has more.

   If your child answers correctly, point to the larger pile and tell your child, "Yes, that pile has a greater amount." If your child points to the wrong pile, count the straws with your child.

3. Ask your child to point to the pile with the lesser amount.

4. Repeat this process for a few more piles of straws.

   You can create two piles with the same number of straws and tell your child that the piles have an "equal" number of straws.

TIP

Repeat this process with other objects, such as two stacks of pennies, stacks of silverware, and so on. Working with different objects will solidify your child's understanding of greater than, less than, and equal.

Comparing numbers on a number line

At this point, your child should be familiar with number lines and counting. They should also know that number lines normally start with 1 (I get to zero later in Chapter 7) and that they count up from there. Here is a number line with the numbers 1 through 10:

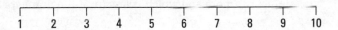

To use the number line to introduce the concept of more-than and less-than, do the following:

1. Explain to your child that the numbers increase, which means that a number to the left is less than the number to its right.

2. Point to the numbers 2 and 3 and say, "The number 3 is greater than the number 2."

3. Circle a number on the number line. Ask your child to identify the numbers that are less than the number and then to point to the numbers that are greater.

Learning the Greater-than, Less-than, and Equal Symbols

Previously, your child learned to compare physical objects to determine which is greater than or less than, or if the objects are the same. In this section, you'll teach your child how to compare numbers and determine which is bigger (greater than), which is smaller (less than), and which are equal.

Learning the less-than symbol

Using your 3x5 index cards, create cards with the symbols greater-than (>), less-than (<), and equal (=).

Locate the cards you created in Chapter 3 with the numbers 1 through 20, and have them handy along with the three cards you just created. Then follow these steps:

TIP

As your child compares two numbers, you may want to place the corresponding number of straws under each number for your child to count to determine which number is more and which is less (or if the numbers are the same).

1. Show your child the less-than symbol and say, "This is the less-than symbol. We use it to show that one number is smaller than another number."

2. Place the cards with 1 and 3 in front of your child with space between them for a third card:

1		3

3. Tell your child that the number 1 is smaller, or less, than 3, and place the card with the less-than symbol between the cards:

1	<	3

4. Move the first set of cards to the side and replace them with the 2 and 5 cards:

2		5

5. Say "2 is less than 5," and have your child place the less-than symbol between the cards:

2	<	5

6. Move the second set of cards to the side and place the following cards in front of your child:

7		5

7. Ask your child, "Is 7 less than 5?"

If your child says "Yes," compare the two numbers on the number line to show your child that 7 is not less than 5.

TIP

Understanding the relationship between numbers is an abstract process. Do not move onto the greater-than symbol until your child understands the concept of less than.

Learning the greater-than symbol

In the previous section, your child learned to identify whether one number is less than another. In this section, you reverse the process, identifying the larger number that is greater than the other, as described in these steps:

1. **Present the greater-than symbol to your child and say, "This is a greater-than symbol. We use it when the first number we are comparing is bigger than the second number."**

2. **Place the 7 and 5 cards on the table in front your child with space for a card between them.**

7		5

3. **Have your child place the greater-than symbol between the cards:**

7	>	5

4. **Move the cards to the side and replace them with the following:**

9		2

5. **Ask your child which number is bigger.**

    If your child answers correctly, follow up by asking, "If the first number is bigger, which symbol do we use?" Have your child place the appropriate symbol between the cards:

9	>	2

    If your child makes a mistake when identifying the larger number, place the appropriate number of straws beneath each number and ask your child to count the straws in each pile to identify which pile has more.

TIP

Point out to your child that the greater-than (>) and less-than (<) symbols look similar to an arrow, and that the point of the arrow will always point to the smaller number.

Learning the equal symbol

To teach your child about using the equal symbol, you'll need a second set of numbers. Use your 3x5 cards to create a second set of cards with the numbers 1 through 10:

1	2	3	4	5	6	7	8	9	10

Using the two sets of identical cards, you can pick two numbers that are the same (equal) for your child to compare. Again, you may want to place straws beneath each number to give your child something to count to confirm the numbers are the same.

1. **Pick up the two cards with the number 4 and place them in front of your child with space for a card between them:**

4		4

2. **Ask your child if the cards are different or the same.**

   When your child says that the numbers are the same, introduce the equal (=) symbol and say, "This is the equal symbol. We use it when the cards are the same."

3. **Have your child place the equal symbol between the cards:**

4	=	4

4. **Repeat this process for two cards with 3 and then two cards with 5.**

Practicing all three symbols

When your child is proficient with using the greater-than, less-than, and equal symbols, it's time to practice with all three symbols:

1. **Hand your child the cards with the greater-than, less-than, and equal symbols.**

2. **Place two cards at random on the table with space for another card between them. Have your child place the appropriate symbol between the cards.**

3. **Continue placing number combinations on the table and asking your child to use one of the symbols to make a comparison.**

TIP

Comparing numbers is a complex process — one on which teachers may practice for a few weeks within the classroom. Repeat this process over several sessions until your child masters the symbols.

Comparing numbers using a deck of cards

In Chapter 2, you use a deck of cards to play a game with your child to practice identifying the numbers 1 through 9. Here, you play a similar game but you practice making comparisons.

Open a deck of playing cards and remove the face cards. Then, spread the cards face down on the table. Tell your child that you are going to play a card game. Explain that they will pick up a card and then you will pick up one. If your card is greater, you get to keep both cards. If their card is greater, they get to keep the cards. If the cards are the same, you will put the cards back on the table face down. At the end of the game, the person with the most cards wins!

Comparing numbers on a worksheet

Worksheet 4-1 at www.dummies.com/go/teachingyourkidsnewmathfd shows pairs of numbers your child can compare using the greater-than, less-than, or equal symbols. Remind your child that the greater-than and less-than symbols always point to the smaller number. I've included a couple of problems here that you can help your child complete. Then ask them to complete the rest.

3		7

9		2

Knowing and Counting the Numbers 1 through 30

By now, your child knows the numbers 1 through 20. Your child should also know the relationship between the numbers 1 and 11, 2 and 12, 3 and 13, and so. You will use the relationship between 1, 11, and 21; 2, 12, and 22; and 3, 13, and 23 to help your child learn the numbers 21 through 30. With this knowledge in hand, learning to count to 30 should feel familiar to your child. Follow these steps:

1. **Using your 3x5 cards, create cards with the numbers 1 through 30.**

You also can use the cards for 1 through 20 you previously created and make new cards for 21 through 30.

2. **Review the cards with the numbers 1 through 20 with your child.**

Do not proceed with additional numbers until your child has mastered the numbers 1 through 20.

3. **Place the cards with the numbers 1 through 20 in two rows in front of your child:**

1	2	3	4	5	6	7	8	9	10

11	12	13	14	15	16	17	18	19	20

4. **Introduce the card with the number 21 by saying, "This is 21. It comes after 20."**

Place the number 21 card at the start of row 3:

1	2	3	4	5	6	7	8	9	10

11	12	13	14	15	16	17	18	19	20

21

5. **Have your child count the cards. Then, introduce the card with the number 22.**

Have your child place the number 22 card next to the card with 21. Then have your child count the numbers.

6. **Point out to your child that 1, 11, and 21 each have the number 1 in them.**

7. Repeat steps 4 through 6 for the cards through 29:

1	2	3	4	5	6	7	8	9	10

11	12	13	14	15	16	17	18	19	20

21	22	23	24	25	26	27	28	29

8. Present the card with the number 30 to your child and say, "This is 30. It comes after 29."

Have your child place the card in the correct location.

9. Have your child point to each card, saying the corresponding number out loud.

10. Turn the cards from 1 through 10 face down.

11. Have your child point to each card and say the number, doing the numbers from memory.

12. Turn the cards 11 through 20 face down, and repeat Step 11.

13. Turn cards 21 through 30 face down and repeat Step 11.

If your child pauses, allow them to turn over the card and look.

Repeat this process until your child has mastered the numbers.

Turn the cards face up and mix them up randomly as you spread them across the table. Ask your child to put the cards back in numerical order. If they pause, allow them to use the number line shown here:

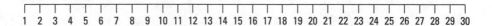

1 2 3 4 5 6 7 8 9 10 11 12 13 14 15 16 17 18 19 20 21 22 23 24 25 26 27 28 29 30

Filling in the gaps: Identifying missing numbers from 1 through 30

A key skill for your child is recognizing and being able to say numbers as they look at them. In this section, you raise the bar by having them identify the numbers even when they aren't there.

Worksheet 4-2 at www.dummies.com/go/teachingyourkidsnewmathfd shows a few number lines from 1 through 30 with missing numbers. Help your child fill in the first few missing numbers and then ask them to complete the rest.

Comparing numbers from 1 through 30

Earlier in this chapter, your child learned to compare numbers using the greater-than, less-than, and equal symbols. In this section, they find out how to compare numbers 1 through 30.

Using the 3x5 cards with the symbols >, <, and =, as well as your cards with the numbers 1 through 30, place two cards on the table and have your child place the appropriate symbol between the cards:

12	<	23

Repeat this process with different cards until your child masters the process. Review the concept over several sessions.

TIP

Comparing numbers 1 through 30 using a worksheet

Worksheet 4-3 at Dummies.com contains a worksheet your child can use to compare numbers from 1 through 30 using the greater-than (>), less-than (<), and equal (=) symbols. I've included a couple of examples here. Help your child complete them and then ask your child to work on the rest of the comparisons on the worksheet.

13		7

29		21

Revisiting Counting Straws

Using your index cards with numbers from 1 through 30 and 30 straws, place a card with a number in front of your child. Ask your child to count the corresponding number of straws. Repeat this process for each of the cards.

Chapter **5**

Understanding Positions, Sorts, and Categorization of Objects

The old proverb, "There's a place for everything, and everything in its place," cuts right to the chase because without order, we'd have chaos! In this chapter, you help bring order to your child's life by teaching them how to position, sort, and categorize objects. Along the way, you're also going to teach them to count to fifty — far beyond what they can count on their fingers and toes. So, put your socks and shoes back on, and let's get started!

TIP

For this chapter, you'll need the following supplies:

» Box of 100 straws

» Deck of 3x5 index cards

» Deck of playing cards

Understanding In Front, Behind, and Next To

Before your child learns to sort various objects, they should first learn the concepts of in front, behind, and next to — doing so introduces the concept of ordering To start, stand up with your child, and ask them to stand in front of you. Explain to your child that they are in front of you and that you are behind them. Step beside your child and say, "Now, I am next to you!" When your child seems comfortable with these concepts as they apply to their position in the environment, you can move on to discussing the spatial relationship of various objects to each other.

Take two different objects, such as your deck of cards and a few straws. Ask your child to put the deck of cards in front of the straws. Then ask them to move the cards next to the straws. Finally, have your child place the cards behind the straws.

Categorizing Objects

Categorizing is the process of selecting items that correspond to a specific group. Your child will use the ability to categorize objects throughout their lifetime. There are many attributes you can use to categorize objects, such as by color, size, shape, and so on.

Use your deck of playing cards to introduce your child to the process of categorizing by following these steps:

1. **Ask your child to create two separate piles: one pile for cards with faces and one pile for cards with numbers.**

   The aces can go with the numbers.

2. **Recombine and shuffle the cards and have your child categorize the cards based on color (red and black).**

3. **Recombine and shuffle the cards again, and then ask your child to categorize the cards by suit.**

   You should have four piles: hearts, clubs, diamonds, and spades.

TIP

Take opportunities throughout the day to talk with your child about categories, and practice separating objects that you encounter. Here are some examples:

>> Separating laundry by type (socks, pants, shirts)

>> Placing silverware into the appropriate location within a drawer

>> Placing groceries in the correct location

>> Grouping objects (such as socks, towels, and washcloths) by color or size

Working up to Fabulous 40

By now, your child probably has mastered the numbers 1 through 30, but it's still a good idea to do a quick review. After a little refresher, you'll move through the thirties to 40. By the end of this chapter, you and your child will be halfway to 100!

For the review, you can reuse the cards you previously created for numbers 1 through 30. If some of them are getting a little tattered, you can replace them with new ones. Also create new cards for numbers 31 to 40.

Use the following steps to review numbers 1 through 30 with your child. Do not proceed with teaching additional numbers until your child has mastered the numbers 1 through 30.

1. Place the cards with the numbers 1 through 30 in three rows in front of your child.

2. Introduce the card with the number 31 and say, "This is 31. It comes after 30."

Start a fourth row with the 31 card.

1	2	3	4	5	6	7	8	9	10
11	12	13	14	15	16	17	18	19	20
21	22	23	24	25	26	27	28	29	30
31									

3. **Have your child count all the cards and then introduce the 32 card.**

Ask your child to place the 32 card next to the 31 card. Then have them count all the numbers again.

4. **Repeat Step 3 for the cards 33 through 39.**

With each additional card, point out the relationships between cards, such as that 1, 11, 21, and 31 each have a 1.

5. **Present the card with the number 40 and say, "This is 40. It comes after 39."**

Have your child place the card in the correct location at the end of the fourth row.

6. **Have your child point to each card as they say the corresponding number.**

7. **Turn cards 1 through 10 face down and then ask your child to point to each card and say the corresponding number, doing the numbers 1 through 10 from memory.**

8. **Turn over cards 11 through 20 and repeat the process.**

9. **Turn over cards 21 through 30 and repeat the process.**

10. **Turn over cards 31 through 40 and repeat the process.**

If your child pauses, allow them to turn over the card and look at the number. If your child gets stuck, celebrate the numbers they have learned and take a break for the day.

TIP

Repeat this process until your child has mastered the numbers.

As a last activity related to counting the numbers in order, turn all the cards face up on the table but mix them up randomly. Ask your child to put the cards back in numerical order. If they pause, allow them to use the number line shown here:

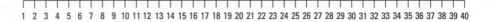

1 2 3 4 5 6 7 8 9 10 11 12 13 14 15 16 17 18 19 20 21 22 23 24 25 26 27 28 29 30 31 32 33 34 35 36 37 38 39 40

Identifying Missing Numbers from 1 through 40

For practice, I've provided a series of number lines spanning 1 through 40 but with some numbers missing. Explain to your child that they're going to work on identifying the missing numbers. You can let your child know that it's okay to start counting from 1 if they need to do that to work up to the missing number. Help your child fill in the first few missing numbers on the following number lines and then ask them to complete the rest.

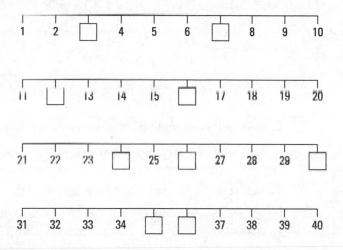

Moving on to Fantastic 50

Your child is learning a lot of numbers. In fact, they are well on their way to learning to count to 100! Make sure to congratulate them for all their effort and progress.

TIP

The process of learning new numbers may take several days. You may decide it's best to work on numbers 1 through 40 on one day and then introduce 41 through 50 on the next.

Use your 3x5 cards to create cards for numbers 41 through 50.

TIP

Review the cards with numbers 1 through 40 with your child. Do not proceed with additional numbers until they have mastered numbers 1 through 40.

1. **Place the cards with the numbers 1 through 40 in four rows in front of your child.**

2. **Introduce the 41 card and say, "This is 41. It comes after 40."**

   Start a fifth row with the 41 card:

1	2	3	4	5	6	7	8	9	10

11	12	13	14	15	16	17	18	19	20

21	22	23	24	25	26	27	28	29	30

31	32	33	34	35	36	37	38	39	40

41

3. **Have your child count all the cards, and then introduce the 42 card.**

   Ask your child to place the 42 card next to the 41 card. Then have them count all the numbers again.

4. **Repeat Step 3 for the cards 43 through 49.**

5. **Present the card with the number 50 and say, "This is 50. It comes after 49."**

   Have your child place the card in the correct location at the end of the fifth row.

6. **Have your child point to each card as they say the corresponding number.**

7. **Turn cards 1 through 10 face down and then ask your child to point to each card and say the corresponding number, doing the numbers 1 through 10 from memory.**

8. **Turn over cards 11 through 20 and repeat the process.**

9. **Turn over cards 21 through 30 and repeat the process.**

10. **Turn over cards 31 through 40 and repeat the process.**

11. **Turn over cards 41 through 50 and repeat the process.**

    If your child pauses, allow them to turn over the card and look at the number.

TIP

Repeat this process until your child has mastered the numbers.

The big finale for counting the numbers in numerical order is to turn the cards face up on the table but mix them up randomly. Ask your child to put the cards back in order. If your child pauses, allow them to use the number line shown here:

1 2 3 4 5 6 7 8 9 10 11 12 13 14 15 16 17 18 19 20 21 22 23 24 25 26 27 28 29 30 31 32 33 34 35 36 37 38 39 40 41 42 43 44 45 46 47 48 49 50

Identifying Missing Numbers from 1 through 50

I've provided a series of number lines from 1 through 50 with some numbers missing. Tell your child that they're going to work on identifying the missing numbers. Help them fill in the first few missing numbers on the following number lines and then ask them to complete the rest.

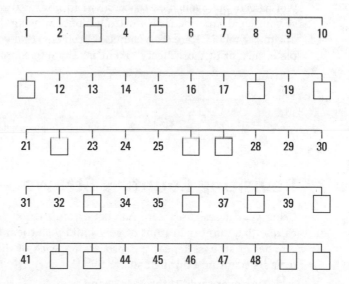

Comparing Numbers from 1 through 50

Using the 3x5 cards with the symbols >, <, and = as well as the cards with the numbers 1 through 50, practice number comparisons with your child.

1. **Hand your child the cards with the greater-than, less-than, and equal symbols.**

2. **Place two cards on the table and have your child place the appropriate symbol between the cards.**

   Here's an example:

22	<	47

3. **Continue placing number combinations on the table and asking your child to use one of the symbols to make a comparison.**

TIP

Repeat this process with different cards until your child masters the process. Review the concept over several sessions.

Comparing Numbers 1 through 50 Using a Worksheet

Worksheets are available at www.dummies.com/go/teachingyourkidsnewmathfd so your child can practice comparing numbers using the greater-than, less-than, or equal symbols. Here are a couple of problems that you can help your child complete; then print Worksheet 5-1 and ask them to complete the rest.

3		7

10		7

9		2

6		16

Revisiting Counting Straws

Using your index cards with numbers from 1 through 50 and 50 straws, place a card with a number in front of your child. Ask them to count the corresponding number of straws. Repeat this process for each of the cards, or until your child masters counting 1 through 50.

IN THIS CHAPTER. . .

» Making size comparisons

» Becoming familiar with tools for measurement

» Using a scale to weigh objects

» Working up to 100

» Understanding the shape of things

Chapter **6**

Getting Specific about Size and Shape

As a great builder (and philosopher) once shared, "Measure twice and cut once!" In this lesson, your child will learn what they need to know to apply this maxim. You're going to break out the ruler and tape measure and teach your child about feet and inches. Your child will also learn to weigh things using a scale. (In case you're wondering, scales and their results haven't changed with the new math.)

TIP

For this chapter, you'll need the following supplies:

» Ruler

» Tape measure

» Bathroom scale

» Flash cards

» Straws

Knowing the Concepts of Bigger, Smaller, and the Same

Before your child learns to measure objects, it is important that they know the concept of bigger, smaller, and the same. A good place to start is with the family socks and shoes, which kids often find it fun to compare. If you have three different-sized shoes — perhaps mom's, dad's, and your child's — that's a great start. If you only have two different-sized shoes, you can make that work, too! Try these steps:

1. Place the shoes in front of your child and say, "These shoes are different sizes. This one is big and this one is small."

2. Ask your child to point at the bigger shoes and then ask them to point at the smaller pair.

3. Place the shoes in a pile and ask your child to put the shoes in a line, from smallest to biggest.

Ask your child to name the shoe's owner as they position them. If you have two paris of shoes for each person, pick up one shoe and ask you child to find the second shoe that is the same.

Learning to Measure

In this lesson, your child will measure different objects using the shoes from the previous exercise. Shoes will work well for this exercise because some adult shoes are about one foot long, which may help your child begin to make sense of the term *foot*. (Although, with that in mind, the metric system may *not* make sense!)

Using one of the biggest shoes and one of the littlest shoes, follow these steps:

1. Put the shoes together, heel to heel, and say to your child, "Let's see how much bigger this shoe is than the smaller one."

2. Place your finger at the toe of the smaller shoe and move the shoe's heel even with your finger, counting "One, two," and possibly "three," as you move the shoe to see how many lengths of the smaller shoe it takes to be equivalent to the larger shoe.

3. Ask your child to measure the big shoe with the little one to see if they agree with you.

4. **Pick an object, such as a table, and ask your child, "How many of your shoes does it take to measure the table?"**

   Help your child measure the table using the shoe by using your finger as a placeholder.

5. **Ask your child to then measure the table using the bigger shoe.**

Learning to measure with a ruler

A shoe may often be easier to come by than a ruler, but it's not the most accurate or standard way to measure something. For that, we have rulers, and your child will use a ruler for a variety of purposes throughout their school years. Use these steps to introduce the concept of measuring with a ruler:

1. **Present the ruler to your child and explain that they can use the ruler to measure things.**

2. **Have your child measure the table using the ruler.**

   Help your child move the ruler across the table. Do not worry about measuring inches yet. You can say something like, "The table is almost four rulers long."

3. **Explain to your child that because the ruler is about the same length as a big shoe, you say the ruler is one foot long.**

   Tell your child that the table, for example, is about 3 feet wide.

4. **Have your child measure other objects, such as a sofa or chair.**

Understanding and measuring inches

Show the ruler to your child and tell them that they will often need to measure objects that are smaller than one foot. Tell your child that to measure small objects, the ruler has smaller marks called *inches*. Use these steps to explain the concept of inches on the ruler:

1. **Point out 1 inch, 2 inches, and 3 inches to your child and show them that there are a total of 12 inches in one foot.**

2. **Using the ruler, help your child measure their shoe.**

   Then find other small objects, such as your phone or a 3x5 index card for your child to measure.

3. **Tell your child that understanding inches helps them to measure items more accurately.**

4. **Have your child use the ruler to measure the table again, counting the feet and then noting the inches.**

   Say something like, "The table is 3 feet and 6 inches wide."

5. **Have your child count out loud as they measure, "One foot, two feet," and so on.**

Drawing lines using the ruler

Using a piece of paper and a pencil, explain to your child that they can use the ruler to draw lines that are a specific number of inches.

1. **Ask your child to draw a line 3 inches long by moving the pencil along the ruler.**

2. **Ask your child to draw a 5-inch line beneath the line they drew in Step 1.**

3. **Ask your child which line is longer.**

   If your child answers correctly, say, "That's right; 5 inches is longer than 3 inches." If they answer incorrectly, say, "Let's measure each line and find out!"

Using a tape measure

Sometimes a big measurement job requires a tool other than a ruler, such as a yardstick or a tape measure. Explain to your child that often we must know the length of a room or furniture and that using a ruler to measure such large objects would be hard. Present the tape measure to your child and show them that, like the ruler, it measures inches and feet.

TIP

Measuring large objects with a tape measure can be tricky. Help your child with the tape measure by holding one end.

1. **Introduce the tape measure to your child by saying, "This is a tape measure; it's much longer than a ruler. We can use it to measure large objects."**

2. **Show your child the feet and inch markings on the tape.**

3. **Use the tape measure to measure different objects.**

4. **Have your child note the number of feet and inches.**

5. **Take time to measure how tall your child is.**

Weighing Different Objects

Length isn't the only type of measurement your child will need to know. They also need to understand weight and the concepts of lighter and heavier. Follow these steps to introduce these concepts:

1. **Have your child hold the ruler in one hand and the tape measure in the other.**

   Ask your child which object is heavier.

2. **Explain to your child that you often want to know how much something weighs and that you use a scale to measure weight.**

3. **Place the scale in front of your child and say, "Let's use the scale to see how much you weigh."**

4. **Have your child stand on the scale and note the number.**

5. **Explain to your child that a scale measures weight using pounds.**

   Tell your child how much they weigh in pounds.

6. **Allow your child to place other objects on the scale and note the different weights.**

7. **Remind your child that when they specify a weight, they should also say "pounds."**

Edging up to 80

In previous lessons, your child has mastered the numbers 1 through 50. Now you will work on the numbers through 80.

Using your 3x5 cards, create cards with the numbers 51 through 60.

TIP

Review the cards with the numbers 1 through 50 with your child. Do not proceed with additional numbers until your child has mastered the numbers 1 through 50.

When you're sure your child has mastered the numbers through 50, create cards for numbers 51 through 60 and use these steps to introduce the next set of numbers:

1. **Place the cards with the numbers 1 through 50 in five rows in front of your child.**

2. Introduce the card with the number 51 by saying, "This is 51. It comes after 50."

Place the card at the start of row 6:

1	2	3	4	5	6	7	8	9	10
11	12	13	14	15	16	17	18	19	20
21	22	23	24	25	26	27	28	29	30
31	32	33	34	35	36	37	38	39	40
41	42	43	44	45	46	47	48	49	50

51

3. Have your child count the cards, and then introduce the 52 card.

Ask your child to place the 52 card next to the 51 card. Then have your child count all the numbers again.

4. Repeat this process for the cards through 59 (you should have six rows of 10 cards).

With each additional card, point out the relationships between cards, such as that 1, 11, 21, 31, 41, and 51 each have a 1.

5. Present the card with the number 60 to your child and say, "This is 60. It comes after 59."

Have your child place the card in the correct location.

6. Have your child point to each card as they say the corresponding number.

7. Turn cards 1 through 10 face down and then ask your child to point to each card and say the corresponding number, doing the numbers 1 through 10 from memory.

8. Turn over cards 11 through 20 and repeat the process.

9. Turn over cards 21 through 30 and repeat the process.

10. Turn over cards 31 through 40 and repeat the process.

11. Turn over cards 41 through 50 and repeat the process.

12. **Turn over cards 51 through 60 and repeat the process.**

If your child pauses, allow them to turn over the card and look at the number. Repeat this process until your child has mastered the numbers.

Create flash cards for 61 through 80 and repeat these steps to introduce the numbers 61 through 70 and 71 through 80.

Turn all the cards face up on the table and mix them up randomly. Ask your child to put the cards back in numerical order. If they pause, allow them to use the number line shown here:

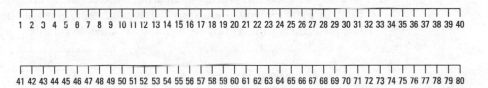

1 2 3 4 5 6 7 8 9 10 11 12 13 14 15 16 17 18 19 20 21 22 23 24 25 26 27 28 29 30 31 32 33 34 35 36 37 38 39 40

41 42 43 44 45 46 47 48 49 50 51 52 53 54 55 56 57 58 59 60 61 62 63 64 65 66 67 68 69 70 71 72 73 74 75 76 77 78 79 80

Identifying missing numbers from 1 through 80

Worksheet 6-1 at www.dummies.com/go/teachingyourkidsnewmathfd shows a series of number lines spanning 1 through 80 but with some numbers missing. Explain to your child that they're going to work on identifying the missing numbers. Help your child fill in the first few missing numbers, and then ask them to complete the rest. You can let your child know that it's okay to start counting from 1 if they need to do that to work up to the missing number, or that they can use the number line in Figure 6-1.

Comparing numbers from 1 through 80

Using the 3x5 cards with the symbols >, <, and = as well as your cards with the numbers 1 through 80, practice number comparisons with your child:

1. **Hand your child the cards with the greater-than, less-than, and equal symbols.**

2. **Place two cards on the table and have your child place the appropriate symbol between the cards.**

Here's an example:

| 63 | < | 77 |

3. **Continue placing number combinations on the table and asking your child to use one of the symbols to make a comparison.**

TIP

Repeat this process with different cards until your child masters the process. Review the concept over several sessions.

Comparing numbers through 80 using a worksheet

Worksheets are available at www.dummies.com/go/teachingyourkidsnewmathfd so that your child can practice comparing numbers using the greater-than, less-than, or equal symbols. Here are a couple of problems you can help your child to complete; then print Worksheet 6-2 and ask them to complete the rest.

3		7

33		33

51		46

79		80

Counting straws through 80

Using your index cards with numbers from 1 through 80 and 80 straws, place a card with a number in front of your child. Ask your child to count and give to you the corresponding number of straws. Repeat this process for several cards to ensure that your child has mastered the skill.

Evaluating Common Shapes

A key kindergarten requirement is that your child can recognize common shapes. Figure 6-1 contains shapes your child must learn to identify.

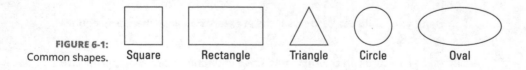

FIGURE 6-1:
Common shapes. Square   Rectangle   Triangle   Circle   Oval

Introduce each shape to your child and ask them to repeat the shape's name. As you introduce a shape, point out different attributes about each one:

>> A square has four equal sides.

>> A rectangle also has four sides, but all four are not the same. Show your child that the two opposite sides of the rectangle are equal.

>> A circle does not have any sides; it is round.

>> A triangle has three sides.

>> An oval is like a circle and does not have sides, but the oval is elongated, like someone stretched it.

Using your 3x5 index cards, draw the shapes on the cards. Then use the cards as flash cards, presenting one card at a time to your child. After your child knows the shapes, place the cards on a table, name a shape, and ask your child to pick up the corresponding card.

TIP

If you have marshmallows, a fun activity is to have your child push the marshmallows onto the end of the straws to build the square and triangle.

Figure 6-2 shows the common shapes with space for your child to trace and then draw them. Using a pencil, help your child draw the shapes.

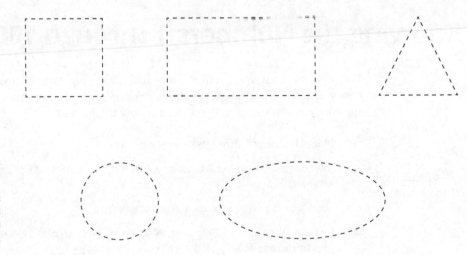

FIGURE 6-2:
Tracing and drawing common shapes.

Shapes can have more than two dimensions, and this section is designed to help your child understand how flat drawings of shapes on the page relate to shapes in the real world. Figure 6-3 shows advanced shapes: the sphere, cube, and pyramid.

FIGURE 6-3: Advanced shapes.

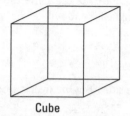  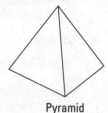

Cube   Sphere   Pyramid

Present the shapes to your child. With each shape, point out key attributes, such as the following:

>> A cube looks like multiple squares.

>> A sphere is round like a ball.

>> A pyramid looks like multiple triangles.

Ask your child to name each shape.

Knowing the Numbers 1 through 100

Your child is about to learn to count to 100! That's big! You will use the same process you used in the "Edging up to 80" section earlier in this chapter. Create cards for numbers 81 through 100 and then follow these steps:

1. **Place the cards 1 through 80 in eight rows.**

2. **Introduce the number 81 and say, "This is the number 81. It comes after 80."**

   Have your child place the card in the correct row.

3. **Introduce the cards 82 through 90 and ask them to place each card in the appropriate spot.**

4. **Repeat Steps 2 and 3 for the numbers 91 through 100.**

5. **Ask your child to point to each card and say the corresponding number.**

6. **Turn cards 1 through 10 face down on the table and ask your child to point to and say each number, doing the numbers 1 through 10 from memory.**

7. **Repeat this process until you've worked through each row.**

Turn the cards face up on the table and mix them up randomly. Ask your child to put the cards back in numerical order. If your child pauses, allow them to use the number line shown here:

Identifying missing numbers from 1 through 100

Worksheet 6-3 at www.dummies.com/go/teachingyourkidsnewmathfd contains a series of number lines from 1 through 100 with some numbers missing. Tell your child they're going to work on identifying the missing numbers. Help your child fill in the first few missing numbers and then ask them to complete the rest. As before, if your child has trouble with a number, help them find the number using the number line.

Comparing numbers from 1 through 100

Using the 3x5 cards with the symbols >, <, and = as well as your cards with the numbers 1 through 100, practice number comparisons with your child:

1. **Hand your child the cards with the greater-than, less-than, and equal symbols.**

2. **Place two cards on the table and have your child place the appropriate symbol between the cards.**

   Here's an example:

   | 18 | < | 100 |

3. **Continue placing number combinations on the table and asking your child to use one of the symbols to make a comparison.**

TIP

Repeat this process with different cards until your child masters the process. Review the concept over several sessions.

Comparing numbers through 100 using a worksheet

Worksheets are available at www.dummies.com/go/teachingyourkidsnewmathfd so that your child can practice comparing numbers using the greater-than, less-than, or equal symbols. Here are a few problems you can help your child to complete; then print Worksheet 6-4 and ask them to complete the rest.

33		37

10		17

99		82

84		84

Chapter **7**

Adding and Subtracting the Numbers 1 through 10

After your kiddo has a good understanding of numbers and how they relate to each other, you can begin teaching them the third of the three r's: readin', writin', and 'rithmetic (though no one ever calls math *arithmetic* anymore).

This chapter helps you introduce the most basic operations in math: addition and subtraction. You start with adding the numbers 1 through 10, move on to subtraction, and then mix up the math problems your child will do. Finally, you will end with a whole bunch of nothing: the concept of zero.

If this chapter seems longer than many others, it's because there's a lot to learning addition and subtraction. But don't worry; your child will master each concept one at a time. Plan to spend a week or more working on this chapter's content.

Before you get started, you'll want to have the following supplies on hand:

» Straws for counting

» Index cards

Starting with Basic Addition

Addition is an essential skill that your child will use throughout their life. To start the process of teaching your child addition, you will work with your straws (or pennies). With the straws as a tangible tool for visualization, your child should quickly grasp the concept of addition. This section explains how to teach your child to add numbers up to 10.

Using straws to understand addition

It's time to pull out the straws, or whatever your kid's counting device is, because you're going to use them to explain the idea of addition. There's no better way for kids to learn than to see concepts in action.

Take out 10 straws and follow these steps to demonstrate the concept of addition:

1. **Give your child 1 straw and ask them how many straws they have.**

2. **Give your child 2 more straws and say, "You had 1 straw, and I gave you 2 more."**

   Ask your child to count the straws. Then ask how many straws they have now.

3. **Say to your child, "1 straw plus 2 straws equals 3 straws."**

4. **Give your child 2 more straws and say, "You had 3 straws, and I gave you 2 more."**

   Ask your child to count how many straws they now have.

5. **If they answer correctly, say, "Yes! 3 straws plus 2 straws equals 5 straws."**

   Tell your child, "You are doing addition!" If your child doesn't say 5, then have them count their straws. Repeat the process several times until they understand the process of adding straws and counting the result.

Practicing written addition using index cards

Now it's time to introduce written addition to your child. Using your 3x5 index cards, find the cards with the numbers 1 through 10. Then create two additional cards: one card with the plus sign (+) and a second card with the equal sign (=):

Now, follow these steps for the next lesson:

1. **Place the card with the number 1 on the table and place 1 straw above it.**

2. **Introduce the card with the plus sign.**

   Explain that this is a plus sign and that it is used for addition.

3. **Place the card with the plus sign next to the 1 card:**

1	+

4. **Place the 2 card on the other side of the plus sign, place 2 straws above the card, and place the equal sign next to the card:**

1	+	2	=

5. **Explain to your child that you add the numbers 1 and 2. Then, have them count the straws and tell you the count.**

6. **Place the number 3 next to the equal sign to complete the equation:**

1	+	2	–	3

   Point to each card and say, "1 plus 2 equals 3."

7. **Place the cards 2 + 3 = on the table, with 2 straws above the 2 and 3 straws above the 3:**

2	+	3	=

8. **Have your child count the straws and tell you the count. Place the number 5 to the right of the equal sign:**

2	+	3	=	5

   Say, "2 plus 3 equals 5."

9. **Repeat Steps 1 through 8 for the following expressions:**

   $2 + 4 =$

   $1 + 3 =$

   $4 + 3 =$

   $5 + 4 =$

Demonstrating comprehension of expressions using straws

The term "expression" is a fancy word for a math statement that has two numbers and an operation, such as addition or subtraction. Up to now, *you've* been placing the straws needed to illustrate the numbers for each expression. Now, it's your child's turn to lay out the straws that match the expression that you've created. Follow these steps:

1. **Point out to your child that you have been placing the straws that illustrate each expression, and now *they* have the chance to place the straws.**

2. **Place the following expression on the table using the index cards:**

   $2 + 1 =$

3. **Ask your child to place 2 straws above the card with the number 2 and 1 straw above the card with the number 1.**

4. **Ask your child to add 2 + 1 by counting the straws. Have your child place the card with the number matching the answer next to the equal sign:**

   $2 + 1 = 3$

5. **Repeat this process of having your child place the straws for the following expressions:**

   $5 + 2 =$
   $4 + 4 =$
   $1 + 1 =$
   $3 + 6 =$

Doing addition on a worksheet

Now that your child has practiced addition operations, they are ready to show their skills on an addition worksheet.

Your child can use Worksheet 7-1 (available at www.dummies.com/go/teachingyourkidsnewmathfd) to count and add objects. Print the worksheet and ask your child to complete it. If your child misses a problem, help them by counting the objects.

Using a number line to add numbers

Previously, you've worked with your child on using number lines to understand *numeric sequences*, that is, which numbers precede or follow others. In this lesson, you show your child how to add with a number line — a technique they will use throughout this book to add even bigger numbers.

The following illustration shows a number line with the numbers 1 through 20.

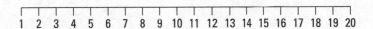

Remind your child that they have used a number line before. Ask your child to count the numbers on the line. Explain to them that they can use a number line to add numbers, and show them examples using the following steps.

1. **Using your index cards, place the following equation on the table:**

   Explain to your child that the first number, 3, tells them where to start on the number line.

2. **Ask your child to find the number 3 on the number line.**

3. **Tell your child that the second number, 4, tells them how much to add to the first number.**

4. **Using the number line, start at the number 3 and then move right as you count, "1, 2, 3, 4," and point to each consecutive number on the number line.**

5. **When you reach 7 on the number line, say, "7, 3 + 4 equals 7."**

6. **Repeat the steps for the following expressions:**

   $3 + 2 =$
   $4 + 4 =$
   $5 + 1 =$
   $6 + 3 =$

Adding numbers on a worksheet using a number line

Your child can use the following number line your child to complete the addition problems in Worksheet 7–2. Help your child solve the first few problems and then ask them to complete the rest.

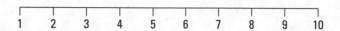

Adding numbers using flash cards

Mastering addition takes practice, and flash cards are one of the best tools for practice. In this section, you create flash cards your child can use to practice addition. Work with the flash cards on daily basis with your child until they can easily solve them.

1. **Using your 3x5 index cards, create the following flash cards:**

1+1	1+2	1+3	1+4	1+5	1+6	1+7	1+8	1+9

2. **Place 10 straws where your child can reach them.**

   As you set out the straws, allow your child to count them.

3. **Beginning with the 1 + 1 card, ask your child to pick up 1 straw and then a second straw, and ask them, "What is 1 + 1?"**

   If your child is correct, say "That's right!" If they are wrong, have them count the straws.

4. **Show your child the card with 1 + 2 and again ask them to pick up appropriate straws and tell you what 1 + 2 is.**

5. **Repeat this process for the 1 + 3 card, reviewing the previous cards as well.**

6. **Repeat this process for the remaining cards, adding one card at a time and reviewing the previous cards.**

TIP

Practice these steps with these cards for several lessons. When your child seems comfortable with going through these addition problems in order, you can mix up the cards so your child can review them when they're out of order. Do not move on to adding numbers to 2 until your child has mastered these cards.

For now, we're only using cards that add up to 10 or less, so in the following steps, you and your child will work on adding with 2, up through 2 + 8:

1. **Using your 3x5 index cards, create the following flash cards:**

2+1	2+2	2+3	2+4	2+5	2+6	2+7	2+8

2. **Introduce the 2 + 1 card and allow your child to count straws to come up with the answer.**

3. **Introduce the 2 + 2 card, repeating the process.**

4. **Show your child the 2 + 1 and 2 + 2 cards again, and then introduce the 2 + 3 card.**

5. **Continue this process to introduce the cards through 2 + 8.**

6. **Combine the 2 + cards with the 1 + cards. For now, keep the cards in order, with 1+1 first and 2+8 last. After you child has mastered the cards in order, you can mix the cards up.**

7. **Review all the cards with your child.**

TIP

Practice with this set of cards until your child has mastered the ones and twos.

Using your 3x5 index cards, create the following flash cards to continue practicing the rest of the number combinations that add up to 10 or less. You will start each lesson by reviewing the cards that you've previously covered with your child. Do not move on to new numbers until your child has mastered the cards. Then introduce the next set of cards.

3+1	3+2	3+3	3+4	3+5	3+6	3+7

4+1	4+2	4+3	4+4	4+5	4+6

5+1	5+2	5+3	5+4	5+5

6+1	6+2	6+3	6+4

7+1	7+2	7+3

8+1	8+2

9+1

Practicing addition through 10 using a worksheet

You may have heard it said that there's no crying in baseball; likewise, there are often no straws in the classroom. Instead, students need to show their understanding on worksheets. To prepare your child for that moment, Worksheet 7-3 at www.dummies.com/go/teachingyourkidsnewmathfd includes expressions your child can use to add numbers through 10. I've included the first few problems here so you can help your child solve them before you turn them loose on the rest of the worksheet:

$$1+1=\underline{\quad} \qquad 2+1=\underline{\quad} \qquad 3+1=\underline{\quad}$$
$$2+2=\underline{\quad} \qquad 2+3=\underline{\quad} \qquad 2+4=\underline{\quad}$$

Take It Away! Understanding Subtraction

Most kids like the idea of adding things: more candy to the pile, more toys in their bedroom, more minutes with their video game. But try to take away something, and hoo boy! Things can get ugly — unless you're taking vegetables off their dinner plate.

In this section, I explain how to teach your child how to subtract numbers up to 10.

Subtracting with straws

Just as you can explain the concept of addition to your child using straws, you can use straws to explain subtraction. You teach subtraction by giving your child a group of straws and then asking them to give some back. (They'll probably be more willing to give up straws than pieces of candy.)

Follow these steps to start introducing the idea of subtraction:

1. Explain what subtraction is to your child.

For example, you can explain that subtraction is the opposite of addition. With addition, you combine (add) two numbers to get a bigger number. With subtraction, you start with a big number and take some away to produce a smaller number. Tell your child that they are going to learn how to subtract two numbers.

2. **Place 3 straws in front of your child and have them count the straws.**

3. **Say to your child, "You have 3 straws. Give 1 to me. How many do you have left?"**

   If your child is correct, say "That's right!" If your child is wrong, count the straws they have in their hand.

4. **Repeat this process with other combinations of straws.**

Practicing written subtraction using index cards

Using the cards you've created for numbers 1 through 10 and the equal sign, create a card with the minus sign:

1. **Place the following expression in front of your child and explain that the minus sign is the subtraction operator. Place 3 straws above the 3 card.**

3		1	–	2

2. **Say to your child, "3 – 1 = 2."**

   Explain to your child that they can also read the expression as "3 take away 1 equals 2."

3. **Take 1 straw away from the pile of 3 straws and have your child count how many straws are left.**

4. **Hand your child 5 straws. Place the following expression in front of your child:**

5	–	3	=

5. **Ask your child to give you 3 straws and count how many straws they have left.**

   Place the number 2 at the right of the expression and say, "5 – 3 = 2":

5	–	3	=	2

6. Repeat Steps 1 through 5 for the following expressions:

$6 - 4 =$
$5 - 1 =$
$8 - 3 =$
$9 - 7 =$
$10 - 2 =$

Learning to subtract using a number line

Tell your child that they can use a number line for subtraction just as they did for addition. Follow these steps to practice with the number line provided as help for Worksheet 7-2.

1. Using your index cards, place the following equation on the table:

7	–	4	=

2. Ask your child to find the number 7 on the number line.

Explain to your child that the first number, 7, tells them where to start on the number line.

3. Tell your child that the second number in the expression, 4, tells them how much to subtract or take away from the first number.

4. Using the number line, start at the number 7 and then move left as you count, "1, 2, 3, 4," and point to each consecutive number on the number line.

5. When you reach the number 3, say, "3, 7 – 4 equals 3."

6. Repeat this process for the following expressions:

$9 - 3 =$
$5 - 2 =$
$10 - 4 =$
$8 - 5 =$

Subtracting numbers on a worksheet using a number line

Worksheet 7-4 at Dummies.com provides exercises your child can do to practice with subtracting numbers. (You can also refer to the number line provided as help for Worksheet 7-2 as you do this.) Help your child solve the first few problems and then ask them to complete the rest. If your child misses a problem, help them to solve the problem using the number line.

Subtracting numbers using flash cards

Whoever said, "Practice makes perfect," must have had flash cards in mind! There's no better way to perfect subtraction skills than by repeated practice with flash cards. Follow these steps for this practice.

1. **Using your 3x5 index cards, create the following flash cards:**

2–1	3–1	4–1	5–1	6–1	7–1	8–1	9–1

2. **Place 10 straws where your child can reach them.**

   As you set out the straws, allow your child to count them.

3. **Begin with the 2 – 1 card and ask your child to pick up 2 straws and then give one to you. Ask your child, "What is 2 – 1?"**

4. **Show your child the card with 3 – 1, allowing your child to pick up three straws. Have your child give one straw back to you. Ask your child what is 3 – 1.**

   If they are correct, "Say yes! 3 – 1 = 2." If they are wrong, have them count the number of straws in their hand.

5. **Go back to the 2 – 1 card and then the 3 – 1 card.**

6. **Repeat Steps 1 through 4 for the 4 – 1 card, repeating the previous cards.**

7. **Repeat this process for the remaining cards, adding one card at a time and reviewing the previous cards.**

TIP

Practice this process with these cards for several lessons. When your child seems to have a grasp on doing the subtraction in numerical order, mix up the cards so they are in random order and review the cards with your child. Do not move on to subtracting 2 until your child has mastered these cards.

When your child has achieved mastery of subtracting 1, create the following flash cards:

3–2	4–2	5–2	6–2	7–2	8–2	9–2	10–2

Then use the following steps to work on problems in which you subtract 2.

1. **Introduce the 3 – 2 card and allow your child to count straws. Hand your child 3 straws and ask them to give 2 straws back. Ask your child, what 3 – 2 is.**

2. **Introduce the 4 – 2 card, repeating the process.**

   Show your child the 3 – 2 and 4 – 2 cards again.

3. **Introduce the 5 – 2 card.**

   Continue this process to introduce the cards up through 10 – 2.

4. **Combine the 2 cards with the 1 cards, and review all the cards with your child.**

TIP

Practice the combined set of cards until your child has mastered the ones and twos.

Using your 3x5 index cards, create the following flash cards:

4–3	5–3	6–3	7–3	8–3	9–3	10–3

5–4	6–4	7–4	8–4	9–4	10–4

6–5	7–5	8–5	9–5	10–5

7–6	8–6	9–6	10–6

8–7	9–7	10–7

9–8	10–8

10–9

REMEMBER

Start each lesson by reviewing the cards your child knows. Do not move on to new numbers until your child has mastered the preceding sets of cards.

Practicing subtraction through 10 using a worksheet

Worksheet 7-5 at www.dummies.com/go/teachingyourkidsnewmathfd includes expressions that your child can use to subtract numbers through 10. I've included the first few problems here so you can help your child solve them before you turn them loose on the rest of the worksheet:

$2-1=$ ___ $3-2=$ ___ $3-1=$ ___

$4-3=$ ___ $4-2=$ ___ $4-1=$ ___

Mixing It Up: Doing Addition and Subtraction

Often, we need to perform addition and subtraction operations together. You might, for example, stop to buy Girl Scout cookies. You need to add up the price of your cookie choices and then subtract that from the amount of money you have to know how much change you will receive.

After your child has mastered the addition flash cards and, separately, the subtraction flash cards, mix the cards together and have your child practice each.

TIP

At first, the process of mixing addition and subtraction problems may seem hard. Take your time. You may have to remind your child which skill they need to use to solve the equation on a card. Be patient. The combined process is a thinking and memorization task. After your child has practiced enough to master it, they will have a strong foundation for moving forward with higher-level math skills.

Worksheet 7-6 at Dummies.com contains exercises that mix addition problems with subtraction problems. Help your child solve the first few, which I've included here, and then ask them to solve the rest.

$2+1=$ ___ $3-2=$ ___ $3+1=$ ___

$4-2=$ ___ $4+2=$ ___ $4-1=$ ___

Understanding the Concept of Zero

If you were wondering why I waited until now to introduce zero, it's because the concept of nothing (Seinfeld's show about nothing excluded), can be a difficult concept to grasp. If you don't believe that, try having your child explain what they mean when they tell you that they've been doing nothing!

Create a flash card for 0 and then explain to your child that the number 0 means none. Give your child two straws. Ask your child to give the straws back to you and then ask, "How many straws do you have left?"

The next step is to examine a number line with the numbers 0 through 10, such as the one shown here. Tell your child that 0 is less than 1. Have them point to the zero.

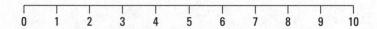

Adding zero to a number

Now that your child understands the concept of 0, they are ready to learn how to add 0 to a number. Use these steps to demonstrate the concept:

1. **Give your child 3 straws.**

2. **Use the flash cards to place the following expression in front of your child:**

3	+	0	=

3. **Ask, "How many straws do you have?" Then say, "If I give you zero more straws, how many will you have?"**

   Explain to your child that the expression is the same as asking, "If you have 3 straws and you add 0 more, how many will you have?"

4. **Repeat this process for the following expressions:**

   $1 + 0 =$
   $4 + 0 =$

Subtracting zero from a number

After your child understands that adding 0 to a number results in the number, they are ready to learn how to subtract 0 from a number. Follow these steps:

1. Give your child 3 straws.

2. Place the following expression in front of your child:

| 3 | – | 0 | = |

3. Ask your child to give you back 0 straws and ask, "How many straws do you have left?"

If your child answers correctly, say, "Yes, 3 – 0 = 3." If your child is wrong, repeat the process, counting the number of straws they have.

4. Repeat this process for the following expressions:

$1 - 0 =$
$4 - 0 =$

Using flash cards to add zero to a number

In this section, you continue practicing with your child to add zero to a number, this time working all the way through flash cards up to 10 + 0:

1. Place the cards for the expression "1 + 0 =" in front of your child:

| 1 | + | 0 | = |

2. Give your child 1 straw and say, "If you have 1 straw and I give you 0 more, how many straws will you have?"

If they answer correctly, say, "Yes, 1 + 0 = 1." If your child is wrong, remind them that adding zero to any number results in the number.

3. Place the following expression in front of your child:

| 2 | + | 0 | = |

4. Give your child 2 straws and say, "If I give you 0 more straws, how many will you have?"

Affirm a correct answer or offer a correction for a wrong answer, as needed.

5. Repeat this process for the remaining numbers through 10.

TIP

When your child has mastered these equations, create equation flash cards and mix them in with the other addition flash cards.

Adding zero to a number on a worksheet

Worksheet 7-7 at Dummies.com includes expressions your child can use to practice adding 0 to numbers through 10. I've included the first few problems here so you can help your child solve them before you turn them loose on the rest of the worksheet:

$$2 + 0 = \underline{\quad} \quad 3 - 0 = \underline{\quad} \quad 3 + 0 = \underline{\quad}$$
$$4 - 0 = \underline{\quad} \quad 4 + 0 = \underline{\quad} \quad 4 - 4 = \underline{\quad}$$

Subtracting zero from a number using flash cards

After your child has mastered adding 0 to a number, they are ready to learn how to subtract 0 from a number using flash cards and working all the way through $10 - 0$. Use the following steps to practice this concept:

1. **Give your child 3 straws.**

2. **Place the following expression in front of your child:**

3	–	0	=

3. **Ask your child to give you back 0 straws and then ask, "How many straws do you have left?"**

   If they answer correctly, say, "Yes, 3 – 0 = 3." If your child is wrong, remind them that subtracting zero from any number results in the number.

4. **Repeat this process for the following expressions:**

   $$1 - 0 =$$
   $$4 - 0 =$$

TIP

When your child has mastered these equations, create equation flash cards and mix them in with the other subtraction flash cards.

Subtracting zero from a number on a worksheet

Worksheet 7-8 at www.dummies.com/go/teachingyourkidsnewmathfd includes expressions your child can use to practice subtracting 0 from numbers through 10. I've included the first few problems here so you can help your child solve them before you turn them loose on the rest of the worksheet.

$2 - 0 = $ ___ $1 - 0 = $ ___ $3 - 0 = $ ___
$4 - 0 = $ ___ $6 - 0 = $ ___ $7 - 0 = $ ___

2

Figuring Out
First Grade Math

Chapter **8**

Adding and Subtracting through 100 without Regrouping

I f you took time to read this chapter's title — yeah, you can look again — you may be asking, "What's regrouping?" It's "old math," baby! The stuff we learned! You know, adding with carrying and subtracting with borrowing — the good stuff. In this chapter, you teach your child how to add and subtract numbers — numbers I've specifically chosen so they don't require carrying and borrowing. Learning to add and subtract is a key concept, and I don't want to muddy the water with regrouping just yet; I'll get there in later chapters. And, as a spoiler alert, many of the new-math techniques don't use regrouping, either.

TIP

Before you get started, you'll want to have a bundle of 150 or more straws on hand.

Identifying the Tens and Ones Places

As your child works with numbers larger than 9, they will first work with tens, then hundreds, and then thousands. Knowing how tens and ones differ is key to your child's success in many math operations.

Use the following steps to introduce the concept of number places to your child:

1. Create ten bundles of 10 straws, binding each with a rubber band; also pull out 9 individual straws.

2. Ask your child to count the individual straws to confirm that there are 9.

3. Point to the bundles of straws and say, "These groups of straws each have 10 straws. You can use them to count by 10 through 100. Let's do that: 10, 20, 30, 40, 50, 60, 70, 80, 90, and 100."

   Hand your child different numbers of straw bundles and have your child count by 10 to come up with the total number of straws.

4. Hand 5 straws to your child and ask them to count them.

5. Give your child 2 more straws and ask, "How many straws do you have now?"

   Remind your child that the individual straws count as 1 straw each.

6. Take back the straws from your child and pick up one bundle of 10 straws.

7. Hand the bundle to your child and say, "10," and then, hand your child 3 straws and say, "11, 12, 13."

8. Take back the straws from your child and then hand them two bundles and 5 individual straws.

9. Tell your child to start with the 10s and to count the straws. Help your child by saying, "10, 20," and allow them to count the individual straws.

10. Repeat this process for the following numbers: 31, 45, 52, 66, and 99.

Recognizing the tens and ones places

Your child will use their knowledge of the tens and ones places as they learn to add and subtract larger numbers. In this section, you will use flash cards to teach them to recognize the tens and ones places. Follow these steps:

1. Create the following flash cards or pull them from your stack of already-made cards:

| 24 | 52 | 45 | 66 | 99 |

2. Place the card with 24 in front of your child and have them say the number.

3. Then say, "You can think of the number 24 as having two parts. The 2 tells you the number of tens and the 4 tells you the number of ones."

4. Place two bundles of 10 straws in front of your child and then set down 4 individual straws.

5. Have your child count the straws, "10, 20, 21, 22, 23, and 24."

6. Repeat this process for the 52 card.

7. Place down the 45 card in front of your child and ask them, "How many groups of 10 do you need?"

8. Help your child count the straws, "10, 20, 30, 40," and then ask, "How many ones do you need?"

   If your child does not know, point to the number 5 on the card and say, "You need 5 straws."

9. Count the straws out loud, "41, 42, 43, 44, and 45."

10. Repeat this process for the cards with 66 and 99.

TIP

Understanding the tens and ones places is key to later adding and subtracting large numbers. Practice this exercise with various numbers until your child masters it. For your child to master the tens-place concept, you may need to practice daily for a week.

Recognizing tens on the number line

The following figure shows a number line from 1 through 100. Present the number line to your child and ask them to point to the tens, from 10 through 100. Say a number, such as 10, 20 or 30 and have your child point to the corresponding number.

1 2 3 4 5 6 7 8 9 10 11 12 13 14 15 16 17 18 19 20 21 22 23 24 25 26 27 28 29 30 31 32 33 34 35 36 37 38 39 40 41 42 43 44 45 46 47 48 49 50

51 52 53 54 55 56 57 58 59 60 61 62 63 64 65 66 67 68 69 70 71 72 73 74 75 76 77 78 79 80 81 82 83 84 85 86 87 88 89 90 91 92 93 94 95 96 97 98 99 100

If your child successfully identifies the tens on the number line, you can work on identifying other numbers on the number line by following these steps:

1. **Hand your child one bundle of 10 straws and ask them to point to the number of straws they have on the number line.**

2. **Hand your child 3 single straws and ask them to count how many straws they have and to point to that number on the number line.**

    Your child may need to count the straws out loud, "11, 12, and 13."

3. **Take back the straws and hand your child two bundles of 10 straws each and 4 singles.**

4. **Ask them to use the number line to show you how many straws they have.**

5. **Repeat this process for the following numbers: 44, 52, and 67.**

Identifying the tens and ones places using a worksheet

Your child has learned to identify tens and ones places within a number and on a number line. In this section, they use a worksheet to decompose numbers into their tens and ones components.

1. **On a piece of paper, write the number 52 with two empty boxes beneath it:**

52	

2. **Explain to your child that they are going to write the number of tens in the first box and the number of ones in the second box.**

5	2

    Help your child write numbers in the boxes:

3. **Repeat this process for the number 47:**

47	

Worksheet 8-1 at www.dummies.com/go/teachingyourkidsnewmathfd contains numbers and boxes into which your child can write the number of tens and ones, just as you did on the sheet of paper in the preceding exercise. Help your child get started and then ask them to complete the rest. By now, your child should have a good understanding of the tens and ones places. To ensure mastery, you might repeat the worksheets for two or three sessions. If your child has difficulty successfully completing the worksheets, take a step back and start again with this chapter's first exercises.

Adding and Subtracting Numbers through 100 without Regrouping

It's time to revisit addition and subtractions, but this time you're going to work with larger numbers than previously. There's no regrouping in this section. First, your child needs to become comfortable with adding and subtracting larger numbers without the extra step of regrouping.

Adding numbers through 100 without regrouping

With the understanding of the tens and ones places under their belt, your child is ready to move on to learning how to add numbers 1 through 100 without regrouping. In other words, the numbers I use in this section can be added without needing to carry a ten.

1. **Create the following flash cards or pull these numbers and signs from your stack of already-made cards:**

11	23	31	40	55	+	=

Also have your bundles of 10 straws and your individual straws available.

2. **Tell your child, "You are going to learn how to add big numbers!" and place the cards for the following equation on the table:**

11	+	23	=

3. **Place one bundle of 10 straws and 1 single straw under the 11 and two bundles of 10 straws and 3 single straws under the 23.**

Ask your child, "If you have 11 straws and you add 23, how many straws do you have?"

4. **Let your child count the straws.**

Start with the bundles of tens and then count the ones.

5. **Repeat this process for the following expressions:**

23	+	31	=
40	+	55	=
55	+	11	=

Using a number line to add large numbers

Just as your child can use a number line to add small numbers, they can use a number line to add large numbers. (Of course, you'll need a longer number line.) Refer to the number line from 1 through 100 earlier in this chapter as you work through the following steps:

1. **Place the following cards in front of your child:**

11	+	13	=

2. **Remind your child that the first number, 11, tells them where to start on the number line.**

Ask your child to find 11 on the number line.

3. **Remind your child that the second number, 13, tells them how much they are adding to 11.**

Explain to your child that to add 13, they will first start with the tens.

4. **Use a pencil to start at the number 11 and add 10 and then 3 ones, as shown here:**

5. Repeat this process for the following expression, using a number line.

14	+	23	=

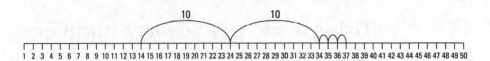

6. Repeat this process for the following expressions, using the number lines shown in the figure:

31	+	12	=
44	+	25	=
61	+	33	=

31 + 12 =

44 + 25 =

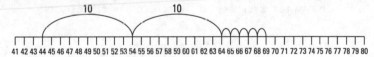

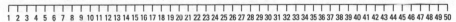

61 + 33 =

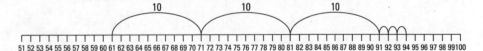

Worksheet 8-2 at this book's companion website at www.dummies.com/go/ teachingyourkidsnewmathfd contains number lines from 1 through 100. Download and print several copies of the PDF. You can use it to create your own addition problems for your child to practice on.

TIP

Using boxes to add large numbers

After your child has mastered using a number line to add numbers, you can teach them to add numbers by using boxes to first add the ones and then the tens. The following problems do not use regrouping, which means your child will not have to carry numbers as they add. (Chapter 15 is where you start to get into that skill.) You will see that the following problems use boxes to reinforce the concept of tens and ones places.

Use the following steps to practice the concept of using boxes for addition:

1. **Present the following boxes to your child:**

	2	3
+	4	4

2. **Tell your child that when adding two-digit numbers, they start by adding the ones column.**

For this example of adding 23 and 44, they will add 3 and 4 first. Have your child write the number 7 in the right column of the bottom row:

	2	3
+	4	4
		7

3. **Tell your child that the second step of adding two-digit numbers is to add the tens column.**

Then ask your child to add the tens column, 2 + 4, writing the result:

	2	3
+	4	4
	6	7

4. **Repeat this process for the following expression:**

3	6
4	1

+

You should get:

3	6
4	1
7	7

+

Worksheet 8-3 at www.dummies.com/go/teachingyourkidsnewmathfd provides addition problems and boxes. Help your child solve the first few and then ask them to complete the rest.

As you perform the addition problems with tens and ones digits, have your straw bundles handy for your child to count as necessary.

TIP

Adding numbers through 100 without regrouping and without boxes

In the previous section, your child solved addition problems that helped them write the tens and ones digits in the correct locations. Now, they will learn to write and solve problems without the boxes.

Present the following problems to your child:

$$\begin{array}{ccccc} 33 & 41 & 55 & 33 & 14 \\ +22 & +13 & +23 & +33 & +21 \end{array}$$

Help your child solve the first problem and then ask them to solve the rest. You should get:

$$\begin{array}{ccccc} 33 & 41 & 55 & 33 & 14 \\ +22 & +13 & +23 & +33 & +21 \\ \hline 55 & 54 & 78 & 66 & 35 \end{array}$$

Worksheet 8-4 on the book's web page has more addition problems that your child can solve. Download it to use for practice.

Do not continue with subtraction until your child has mastered addition.

TIP

Subtracting numbers through 100 without regrouping

Now that your child has an understanding of two-digit addition, they are ready to tackle subtraction of the numbers through 100. Again, I've selected numbers so that these operations don't require regrouping — which in this case means that your child will not need to borrow numbers to complete a subtraction problem.

Here are the steps to introduce subtraction with larger numbers:

1. **With your bundles of 10 straws and your 9 single straws in hand, say to your child, "Now that you know how to add large numbers, you're going to learn how to subtract them!"**

2. **Place 7 straws in front of your child and ask them to count them.**

3. **Take 2 of the straws away and ask your child, "How many straws are left?"**

   If your child answers correctly, say, "Yes, 7 – 2 = 5." If they are wrong, say, "Let's try that again. Start over with the 7 straws, counting them out loud. Then, take away 2 and count the remaining straws out loud.

4. **Pull out the flash cards (or create new cards) for 57, 45, 33, 12, 1, –, and =, and place the following expression on the table:**

33	–	12	=

5. **Using your bundles and single straws, have your child count 33 straws and place them next to the 33 card.**

6. **Ask your child, "How many straws do we need to subtract?"**

   Your child should give you back 12 straws.

7. **Ask, "How many straws are left?"**

   If your child answers correctly, say, "Yes! 33 – 12 = 21." If your child is wrong, repeat the process, helping them by counting the straws out loud as you work.

8. **Repeat this process for the following expressions:**

57	–	45	=
45	–	12	=
57	–	12	=
45	–	1	=

Using a number line to subtract large Numbers

Your child can use a number line to subtract large numbers just as they used it to add large numbers. The following figure shows a number line from 1 through 100, which you can use as you work through the following steps.

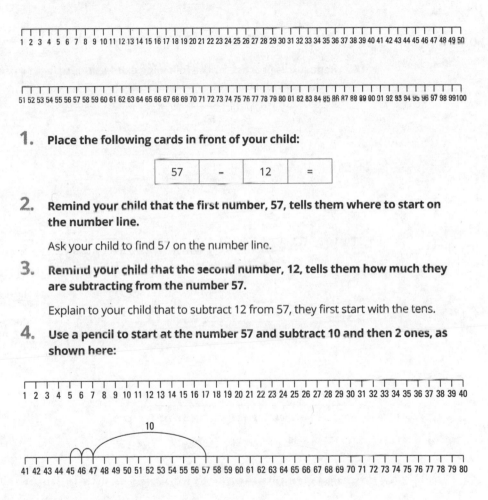

1. **Place the following cards in front of your child:**

57	–	12	=

2. **Remind your child that the first number, 57, tells them where to start on the number line.**

   Ask your child to find 57 on the number line.

3. **Remind your child that the second number, 12, tells them how much they are subtracting from the number 57.**

   Explain to your child that to subtract 12 from 57, they first start with the tens.

4. **Use a pencil to start at the number 57 and subtract 10 and then 2 ones, as shown here:**

5. Repeat this process for the following expression, using the following number line:

| 45 | – | 33 | = |

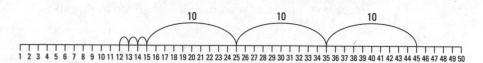

6. Repeat this process for the following expressions using the number lines in the next illustration:

33	–	12	=
44	–	21	=
64	–	33	=

33 – 12 =

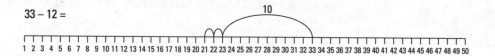

44 – 21 =

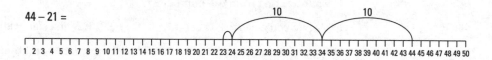

64 – 33 =

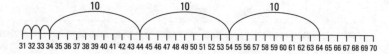

TIP

Download and print several copies of Worksheet 8-2 from the book's companion website. You can use it to create your own subtraction problems for your child to practice on.

Worksheet 8-5, which you can find at www.dummies.com/go/
teachingyourkidsnewmathfd, provides number lines and subtraction problems.
Help your child solve the first few and then ask them to complete the rest.

Using boxes to subtract large numbers

After your child has mastered using a number line to subtract numbers, you can
teach them to subtract numbers by using boxes to first subtract the ones and then
the tens. The following problems do not use regrouping, which means your child
will not have to borrow numbers when subtracting to find the solution. I cover
subtraction with borrowing numbers in Chapter 15.

Use the following steps to practice solving subtraction problems with boxes.

1. **Present the following boxes to your child:**

5	3
4	2

2. **Tell your child that to subtract the numbers, they will start by subtracting the ones column: 3 – 2.**

   Have your child do so and write down the number 1:

5	3
4	2
	1

3. **Ask your child to subtract the tens column, 5 – 4, writing the result:**

5	3
4	2
1	1

4. **Repeat this process for the following expression:**

6	6
3	1

You should get:

	6	6
−	3	1
	3	5

Present the following problems to your child:

```
  47      55      33      14
 −22     −13     −23     −22
```

Worksheet 8-6 at www.dummies.com/go/teachingyourkidsnewmathfd provides subtraction problems and boxes. Help your child solve the first few problems and then ask them to complete the rest.

Subtracting large numbers without regrouping or boxes

The boxes in the preceding section help your child understand how two-digit numbers line up properly in a subtraction problem. Now they can practice the same type of problems without the boxes. Use the following steps:

1. **Present the following problems to your child:**

```
  33      41      55      33      44
 −12     −10     −23     −13     −21
```

2. **Help your child solve the first problem and then ask them to solve the rest. You should get:**

```
  33      41      55      33      44
 −12     −10     −23     −13     −21
  21      31      32      20      13
```

Worksheet 8-7 at the book's web page has more problems that your child can solve. Download it to use for practice.

Mixing addition and subtraction problems on the same worksheet

Now that your child knows how to add and subtract numbers separately, they can begin working addition and subtraction problems on the same worksheet. Worksheet 8-8 at www.dummies.com/go/teachingyourkidsnewmathfd includes both types of problems. Help your child with the first few and then ask them to complete the rest. You may have to remind your child to pay attention to the addition and subtraction signs.

Reviewing Flash Cards for Addition and Subtraction through 10

TIP

Your child's ability to mentally add and subtract numbers quickly is a key skill. One of the best ways to master these operations is to practice with flash cards on a regular basis. The flash cards are not only convenient, but they will increase your child's memorization of key operations.

Use your index cards to create the following addition flash cards:

1 +1	1 +2	1 +3	1 +4	1 +5	1 +6	1 +7	1 +8	1 +9

You also need to make a set of cards for numbers 2 through 9. You will have 81 total addition flash cards when you're done.

TIP

Practice these flash cards daily until your child has mastered them. Do not move on to subtraction until your child has mastered addition.

To practice subtraction, you need another set of cards that look like this:

1 −1	2 −1	3 −1	4 −1	5 −1	6 −1	7 −1	8 −1	9 −1
2 −2	3 −2	4 −2	5 −2	6 −2	7 −2	8 −2	9 −2	
3 −3	4 −3	5 −3	6 −3	7 −3	8 −3	9 −3		
4 −4	5 −4	6 −4	7 −4	8 −4	9 −4			

(continued)

(continued)

5	6	7	8	9
−5	−6	−5	−5	−5

6	7	8	9
−6	−6	−6	−6

7	8	9
−7	−7	−7

8	9
−8	−8

9
−9

TIP

The ability to quickly subtract numbers will provide the foundation for your child's future math success. Practice these flash cards daily until your child has mastered them.

Completing a Timed Worksheet

This book's companion web page at www.dummies.com/go/teachingyourkidsnewmathfd includes Worksheet 8-9, which contains addition and subtraction problems with numbers to 10 for your child to practice. Download and print several copies of the worksheet. Allow your child to practice the problems. Time your child as they work. The goal as your child practices over time is that they should complete the worksheet in five minutes or less.

Chapter **9**

Pairing Numbers with Telling Time

'I've decided that time is a problem; there's never enough of it. What's worse is that time depends on where you live; in many places, for some reason, it somehow falls back and springs forward to save daylight.

If you've ever wondered why Einstein is important, it's because he found that time is relative. Heck, every kid knows that — time goes fast during recess and takes forever when studying math.

In this chapter, you will teach your child about time — specifically, how to tell it.

Before you get started, you'll want to have the following supplies on hand:

» A phone with a clock and stopwatch application

» Analog and digital clocks

Understanding the Concept of Time

Time is an abstract concept. Throughout your day, try to find occurrences when you can note the current time to your child as well as how long different things take (such as recess, preparing dinner, driving to school, or studying math). You can say something like, "Every day, we do things that happen at specific times. You start school at 8:30, you have recess at 10:30, and you eat lunch at 11:30. I pick you up at 3:00 and we eat dinner at 5:30."

Explain that time tells us when something is going to happen, when we must be somewhere, and how long events last. Tell your child that we talk about time using hours and minutes. A favorite TV show, for example, starts at 5:00 and lasts 30 minutes. Recess starts at 10:30 and lasts 10 minutes.

Understanding how time relates to day and night

Even though your child may not yet know how to tell time, they know that certain events occur throughout the day. Tell your child that a day has 24 hours. Explain that for some of the day we are awake, and the rest of the time we're asleep. Explain that when we talk about our day, we use words such as *morning*, *afternoon*, *evening*, and *night*. Using a timeline, write down events in your child's day at the appropriate times. Your timeline might include the following:

>> 7:00 Wake up!

>> 7:30 Eat breakfast

>> 8:00 Leave for school

>> 10:30 Recess

>> 11:30 Lunch

>> 3:00 Come home from school

>> 4:00 Play

>> 5:30 Dinner

>> 7:00 Go to bed

>> 7:30 Sleep!

Here's an example timeline for a 24-hour period. You can use it to mark different activities in your child's day:

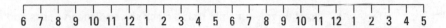

Explain to your child that a day has 24 hours, but a clock only shows 12. That means, in general, 12 hours will happen during the day and 12 hours will happen at night. When we tell someone a time, we will often say, "7:00 in the morning" or "5:30 at night."

Knowing that an hour has 60 minutes

Remind your child that there are 24 hours in a day. Explain that one hour is a long time. Use some examples:

>> At school, you get one hour for lunch and recess.

>> Many movies are two hours long.

>> At night, you sleep for eight hours.

Tell your child that many things take less than one hour. For such things, we use minutes to tell us how long they will last. Again, provide some examples:

>> Your first recess lasts 10 minutes.

>> You practiced math flash cards for 15 minutes.

>> Your TV show lasts 30 minutes.

Also explain to your child there are 60 minutes in one hour.

Identifying hours and minutes on a digital clock

Because there are now digital alarm clocks, digital wrist watches, and even a display on the TV that shows the current time using a digital format, we will start learning to tell time using a digital clock.

Remind your child that one hour has 60 minutes. When they look at a digital clock, they see hours using the numbers 1 through 12 and minutes from 0 to 59.

Figure 9-1 contains digital clocks that show times your child can use to identify hours and minutes. Use these steps to guide your child as you examine the clocks in the figure:

1. **Explain to your child that a clock shows time in hours and minutes.**

   Show your child the hours and minutes on each clock in Figure 9-2 as you say the times out loud.

2. **Tell your child the clock shows hours as numbers 1 through 12 and minutes using numbers 0 through 59.**

TIP

   Using some of the times on the clocks, explain to your child that when the minutes are less than 10, the number of minutes is preceded with a 0, such as 12:05.

FIGURE 9-1:
Hours and minutes on digital clocks.

Adding one minute to a time with 59 minutes

Remind your child that one hour has 60 minutes and that a clock represents the minutes using the numbers 0 through 59. Show your child this clock:

Point to the clock and explain to your child that when the minutes on the clock are at 59 and one more minute passes, the clock increments the hours by 1 and resets the minutes to 0, as shown here:

One minute

Worksheet 9-1 at www.dummies.com/go/teachingyourkidsnewmathfd includes examples your child can use to add 1 minute to times with 59 minutes. Help your child with the first few, and then ask them to complete the rest.

Remind your child that a clock shows hours using the numbers 1 through 12. Explain to your child that when the time is 12:59 and a minute passes, the clock will display the new time as 1:00, as shown here:

One minute

Counting by Fives

At this point, your child should know how to count to 100. To tell time, your child will use the numbers 0 to 59. Confirm that your child knows how many minutes are in an hour by asking, "How many minutes are there in one hour?" You can also ask your child to count out loud to 60, and you can count along with them.

Here is a number line from 1 through 100. Use the number line with the following steps to explain how to count to 100 by fives:

1 2 3 4 5 6 7 8 9 10 11 12 13 14 15 16 17 18 19 20 21 22 23 24 25 26 27 28 29 30 31 32 33 34 35 36 37 38 39 40 41 42 43 44 45 46 47 48 49 50

51 52 53 54 55 56 57 58 59 60 61 62 63 64 65 66 67 68 69 70 71 72 73 74 75 76 77 78 79 80 81 82 83 84 85 86 87 88 89 90 91 92 93 94 95 96 97 98 99 100

1. Say to your child, "You know how to count to 100 by ones. Let's learn to count to 100 by fives."

   Point to the numbers on the number line and count out loud: "5, 10, 15, 20," continuing on until you reach 100.

2. **Pull the flash cards for the following numbers from your stack, or create new cards:**

5	10	15	20	25	30	35	40	45	50
55	60	65	70	75	80	85	90	95	100

3. **Place the cards on the table in front of your child and ask them to say the numbers out loud.**

4. **Turn the 5-card face down:**

	10	15	20	25	30	35	40	45	50
55	60	65	70	75	80	85	90	95	100

5. **Have your child say the numbers, recalling 5 from memory.**

6. **Turn the 10-card face down and have your child say the numbers.**

7. **Repeat this process for the remaining cards.**

Telling Time Using an Analog Clock

You remember analog clocks, right? Your classroom probably had a big one at the front that seemed to slow down as you anxiously watched for the time when recess would start.

Use the following steps to introduce your child to reading time from an analog clock:

1. **Show your child the following clock and say, "This is a different type of clock that also shows the time using hours and minutes."**

2. **Point to the large numbers on the clock to show your child that the clock displays marks for 12 hours.**

3. **Point to each number, saying the number out loud.**

4. **Say, "You know that one hour has 60 minutes. If you look closely at the clock, you will see small tick marks that indicate the minutes."**

5. **Point to the ticks and count out loud from 0 through 59.**

Figure 9-2 shows clocks with different times. Look at the clocks with your child and explain that each clock's little hand points to the hours and the big hand points to the minutes. Follow these steps to practice reading the times on the analog clocks.

6. **Review the time shown on each clock in Figure 9-2 with your child.**

FIGURE 9-2: Clocks with different hours and minutes.

7. Have your child write the times beneath each clock in the hours and minutes boxes.

8. Help your child with the first few, and then ask them to complete the rest.

Knowing the five-minute intervals on the clock

Being able to recognize the five-minute intervals on a clock will help your child quickly tell time. Worksheet 9-2 at www.dummies.com/go/teachingyourkidsnewmathfd includes clocks for practicing five-minute intervals on an analog clock. Use the following steps as you look at the worksheet with your child:

1. Point to a clock on the worksheet and say, "If you look at the minutes on the clock, you will find that each number on the clock occurs at five-minute intervals."

2. Point to each number and say, "This is 5 minutes, 10 minutes, 15 minutes, . . ." through 55 minutes.

3. Have your child complete the worksheet, helping them with the first few.

TIP

At www.dummies.com/go/teachingyourkidsnewmathfd, you can find downloadable Worksheet 9-3, which matches Figure 9-3. After you have downloaded and printed the PDF, do the following:

1. Use scissors to cut out the clock hands.

FIGURE 9-3:
A clock with hour and minute hands.

2. **Remind your child that the small hand points to hours and the big hand points to minutes.**

3. **Place the arrows on the clock face to point them to different times as you ask your child to name the corresponding times.**

TIP

The ability to tell time is important. You may consider putting a clock in your child's room and giving them a watch.

Learning the concept of seconds

Explain to your child that many things happen very quickly, often in less than one minute. Here are several examples:

>> Time to run across the yard

>> Time to fill a glass of water

>> Time to finish a math problem such as 3 + 2 =

>> The length of some commercials on TV

Explain to your child that to time fast events, you use seconds. Tell your child that one second is a very short amount of time and that there are sixty seconds in one minute.

TIP

Most cell phones have a stopwatch app. Open the stopwatch app on your phone and show your child seconds elapsing as the app runs. Teach your child to use the stopwatch app and then allow them to time various events. In addition, most kitchen microwave ovens have a timer that shows minutes and seconds.

Working on Addition and Subtraction through 20

We're getting close to wrapping up second-grade math, and I want to make sure your child has this skill. So I've included a short review section about addition and subtraction.

Reviewing addition through 20

Chapter 7 covers adding numbers through 10. In this section, you will use flash cards to practice adding through 20 — a key skill for successfully performing harder math problems in the future.

For the following exercise, you will need the addition flash cards you created for Chapter 8. You will also need to create some new addition flash cards to add to your set:

1 +10	2 +10	3 +10	4 +10	5 +10	6 +10	7 +10	8 +10
9 +10	10 +10						

TIP

Introduce one row of the flash cards with each lesson until you've worked all the way from 1 + 1 through 10 + 10. Practice the flash cards daily until your child has mastered the cards.

Reviewing addition through 20 using a worksheet

After your child has mastered the flash cards for adding numbers through 20, they are ready to solve similar problems on a worksheet. Worksheet 9-4 at www.dummies.com/go/teachingyourkidsnewmathfd is an addition worksheet through 20. Here are a few problems you can help your child complete before you ask them to complete the rest:

5 + 1

2 + 7

8 + 3

12 + 6

Reviewing subtraction through 20

Just as your child has learned to add numbers through 20, they will now learn to subtract them. Find the subtraction flash cards you used in Chapter 8 and follow these steps to practice subtraction with larger numbers.

Using your 3x5 index cards, create the following flash cards:

20 - 1	20 - 2	20 - 3	20 - 4	20 - 5	20 - 6	20 - 7	20 -8	20 - 9	20 -10
20 -11	20 -12	20 -13	20 -14	20 -15	20 -16	20 -17	20 -18	20 -18	20 -20
19 - 1	19 - 2	19 -3	19 - 4	19 - 5	19 - 6	19 - 7	19 - 8	19 - 9	19 -10

19 −11	19 −12	19 −13	19 −14	19 −15	19 −16	19 −17	19 −18	19 −19	
18 − 1	18 − 2	18 − 3	18 − 4	18 − 5	18 − 6	18 − 7	18 − 8	18 − 9	18 −10
18 −11	18 −12	18 −13	18 −14	18 −15	18 −16	18 −17	18 −18		
17 − 1	17 − 2	17 − 3	17 − 4	17 − 5	17 − 6	17 − 7	17 − 8	17 − 9	17 −10
17 −11	17 −12	17 −13	17 −14	17 −15	17 −16	17 −17			
16 − 1	16 − 2	16 − 3	16 − 4	16 − 5	16 −6	16 − 7	16 − 8	16 − 9	16 −10
16 −11	16 −12	16 −13	16 −14	16 −15	16 −16				
15 − 1	15 − 2	15 − 3	15 − 4	15 − 5	15 − 6	15 − 7	15 − 8	15 − 9	15 −10
15 −11	15 −12	15 −13	15 −14	15 −15					
14 − 1	14 − 2	14 − 3	14 − 4	14 − 5	14 − 6	14 − 7	14 − 8	14 − 9	14 −10
14 −11	14 −12	14 −13	14 −14						
13 − 1	13 − 2	13 − 3	13 − 4	13 − 5	13 − 6	13 − 7	13 − 8	13 −9	13 −10
13 −11	13 −12	13 −13							
12 − 1	12 − 2	12 − 3	12 − 4	12 − 5	12 − 6	12 − 7	12 − 8	12 − 9	12 −10
12 −11	12 −12								
11 − 1	11 − 2	11 − 3	11 − 4	11 − 5	11 − 6	11 − 7	11 − 8	11 − 9	11 −10
11 −11									
10 − 1	10 − 2	10 − 3	10 − 4	10 − 5	10 − 6	10 − 7	10 − 8	10 − 9	10 −10

TIP

Introduce the flash cards, one or two rows per lesson until you've worked all the way from 20 − 1 to 1 − 1. Practice the cards with your child daily until they master the cards.

Reviewing subtraction through 20 using a worksheet

After your child has mastered the subtraction flash cards through 20, they are ready to solve similar problems on a worksheet. Worksheet 9-5 at www.dummies. com/go/teachingyourkidsnewmathfd provides subtraction problems for numbers through 20. Here are a few problems you can help your child complete before setting them loose to complete the rest.

$$4 - 3$$
$$9 - 7$$
$$12 - 6$$
$$15 - 4$$

Counting by 2 through 100

Earlier in this chapter, your child learned to count by fives through 100. Now you'll work on counting by twos, which is useful for understanding the concepts of even and odd, which I introduce in Chapter 21.

Follow these steps to work through this exercise with your child.

1. **Pull the following cards from your stack of flash cards or create new ones for these numbers:**

2	4	6	8	10	12	14	16	18	20
22	24	26	28	30	32	34	36	38	40
42	44	46	48	50	52	54	56	58	60
62	64	66	68	70	72	74	76	78	80
82	84	86	88	90	92	94	96	98	100

2. **Remind your child that they know how to count to 100 by ones and fives, and say, "Now you are going to learn to count to 100 by twos."**

3. Have your child recite the numbers shown on the cards from Step 1.

4. Turn over the first row and have your child say all the cards, completing the first row from memory.

5. Turn over the second row and repeat this process.

6. Continue to turn over rows until your child can easily count to 100 by twos.

Reviewing Counting by 10 through 100

Now that your child has mastered counting to 100 by twos and fives, they are ready to count to 100 by tens. Being able to count by 10 mentally helps your child with addition and subtraction of two-digit numbers using a number line.

Explain to your child that you are going to review counting to 100 by tens. Then use the following steps to do this exercise.

1. Pull the following numbers from your stack of flash cards:

| 10 | 20 | 30 | 40 | 50 | 60 | 70 | 80 | 90 | 100 |

2. Have your child read the numbers on the cards.

3. Turn the 10-card face down and have your child count the numbers, reciting the first one from memory.

4. Turn the 20-card face down and have your child count the numbers, reciting the first two from memory.

5. Repeat this process of turning over cards until your child can easily count to 100 by 10.

Chapter **10**

Getting a Feel for Fractions

Just when math was starting to make sense, someone had to bring in fractions. Everybody uses fractions. You can recognize when your gas tank is less than a quarter full, and you'd happily take half of the pizza instead of one-fourth. You know that Subway's foot-long sandwich is a better buy than two halves. When, however, was the last time you did math with fractions? Sports teams, for example, don't get half a point. All that said, in this chapter, you are going to introduce your child to fractions. The good news is that you won't yet have to add, subtract, multiply, or divide them! That comes later. This chapter's goal is to introduce your child to fractions so that they can recognize them as they encounter them in real life.

Understanding Fractions

When you measure and cut things, you're breaking large pieces into smaller ones. Give your child an example of doing this, such as, "If there are two of us, and we have only one sandwich, I might cut the sandwich in half so you can have one half and I can have the other." (See Figure 10-1.)

FIGURE 10-1:
Slicing a sandwich
in half.

Here's another example you can offer: "If there are four of us, and we have only one pizza, I would slice the pizza into four equal parts." (See Figure 10-2.)

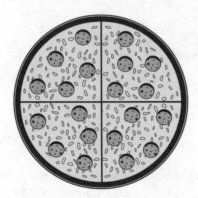

FIGURE 10-2:
Slicing a pizza
into four equal
parts.

After introducing these and other examples of making smaller units from one large unit, work through the following steps to help your child understand the concept of fractions:

1. Explain that when you divide something into smaller pieces, you call those pieces *fractions*.

2. Using the following circle, explain to your child that the circle is whole, meaning it is not divided into smaller pieces.

3. Tell your child that you can divide the circle into two equal parts, as shown here:

4. Explain to your child that when you divide a shape into two equal parts, you refer to each part as a *half*.

5. Point to each part of the circle as you count and say, "One half, two halves."

6. Ask your child to point to each half of the circle.

To reinforce the concept of halves, refer to the following square and follow these steps:

1. Ask your child to help you divide the square in half.

2. Explain to your child that they can draw a line in the middle of the square to divide the square in half, as shown here:

Halves aren't the only way to divide a whole. Use the following steps to explain the idea of thirds to your child.

1. Show your child the following whole circle:

2. Explain to your child that you can divide a whole into three pieces, called *thirds*, and then show them the following divided circle:

3. Explain to your child, "In this case, the circle is divided into three thirds."

4. Point to each third and count, "One third, two thirds, and three thirds."

To reinforce the concept of thirds, refer to the square shown here and follow these steps:

1. Have your child count each third in the following square: "One third, two thirds, three thirds."

Yet another way to divide a whole is into *quarters*, or *fourths*. Referring to the following circle, follow these steps to help your child understand quarters:

1. Tell your child that dividing something, such as the circle, into four equal parts is called dividing it into *fourths*.

Show your child this divided circle:

2. Have your child count each fourth: "One fourth, two fourths, three fourths, four fourths."

To reinforce the concept of creating fourths, refer to the square shown here and follow these steps:

1. Explain to your child that you can divide a square into fourths just as the circle was divided into fourths, and show them this illustration:

2. Have your child count the fourths.

Identifying the Parts of a Fraction

The previous section introduced your child to the concept of fractions. Now, you will teach them how to represent fractions using numbers. Show your child the following circle and use the following steps:

1. Say to your child, "Often, you must represent a fraction using numbers. To do so, you first count the number of parts that are shaded and write that number down with a line beneath it."

2. Ask your child to look at the circle and count how many parts are shaded. Have your child write the number of colored parts with a line beneath it.

3. Tell your child that you also count the total number of pieces and write that number beneath the line.

4. Write the 2 to create the fraction:

$$\frac{1}{2}$$

5. Repeat this process using the following shape:

6. Ask your child to count the shaded parts and write the number with a line beneath, as shown here:

$$\underline{3}$$

7. Ask your child to count the total number of parts in the square in Step 6 and then write the result under the line to complete the fraction as shown here:

$$\frac{3}{4}$$

Worksheet 10-1 at www.dummies.com/go/teachingyourkidsnewmathfd has shapes with shaded portions that your child can use to write the equivalent fractions. Help your child with the first few problems and then ask them to complete the rest on the worksheet. If your child has a problem with any of the fractions, repeat the process of counting and writing down the number of shaded parts above the line and then counting and writing down the total number of parts beneath it.

Drawing Equivalent Fractions

Your child should now be familiar with representing fractions using numbers. Now you are going to reverse the process. You will write a fraction, and they will fill in the appropriate parts of a shape using the following steps:

1. Present the following problem to your child.

2. **Remind your child that the top number of a fraction tells them the number of pieces to shade.**

3. **Use a pencil to shade in the appropriate parts.** The result should look like this:

Repeat these steps for the following shape shown in Figure 10-3; the result is shown in Figure 10-4.

FIGURE 10-3: Dividing a rectangle into portions.

FIGURE 10-4: Shading the portions for five-sixths.

Your child can use Worksheet 10-2 at www.dummies.com/go/teachingyourkids newmathfd to practice this process. Help your child with the first couple of problems and then ask them to complete the rest.

Throughout your day, you may encounter fractions such as when you are baking. When you do, point out and discuss the fraction with your child.

TIP

Understanding Mixed Numbers

Explain to your child that a mixed number has a whole part and a fractional part. Here are some examples:

$$1\frac{1}{2} \quad 2\frac{1}{4} \quad 3\frac{1}{3}$$

Figure 10-5 shows visual representations of these mixed numbers with whole and fractional parts.

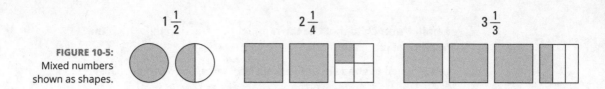

FIGURE 10-5:
Mixed numbers
shown as shapes.

$1\frac{1}{2}$ $\qquad$ $2\frac{1}{4}$ $\qquad$ $3\frac{1}{3}$

Worksheet 10-3 at www.dummies.com/go/teachingyourkidsnewmathfd contains whole and fractional shapes for which your child can write the mixed numbers. Help your child with those problems, and then ask them to complete the rest.

Recognizing Equivalent Fractions

In the world of fractions, the same quantity of a whole can be represented by two different fractions. Show your child the circles in Figure 10-6.

FIGURE 10-6:
A portion of a
circle represented
as one-half and
two-fourths.

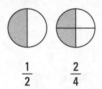

$\dfrac{1}{2}$ $\qquad$ $\dfrac{2}{4}$

Say to your child, "As you can see, we have the fractions one-half and two-fourths, which are different. However, the same amount of each circle is shaded. So, we can say that the fractions one-half and two-fourths are equal."

Repeat the process with the rectangles shown in Figure 10-7.

FIGURE 10-7:
A portion of a
rectangle
represented as
one-half and
three-sixths.

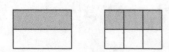

Say, "This time, the two different fractions are one-half and three-sixths. However, the same amount of each shape is shaded. So, we can say that the fractions one-half and three-sixths are equal."

Worksheet 10-4 contains several different fractions with which your child can identify equivalent fractions. Have your child circle the fractions that are equal.

Measuring Using Fractions

Your child has probably used a tape measure and ruler to measure feet and inches. In this section, they will learn that the smaller marks on both tape measures and rulers correspond to fractions of inches. Take out a tape measure or ruler and follow these steps:

1. Say to your child, "When we measure things, we often must use fractions."

2. Show your child the ½-inch marks on the ruler or tape measure.

3. If you are using a ruler, ask your child to draw a line that is 3½ inches long on a piece of paper.

4. Have your child measure different objects in your home with the tape measure.

   Ask them to measure to the half inch.

5. Explain to your child that sometimes you must measure things that are smaller than ½ inch. Show your child the ¼ and ¾ marks on the ruler or tape measure.

6. If you are using a ruler, ask your child to draw a line that is 2¼ inches long.

Comparing Fractions

Your child has learned to compare numbers using the greater-than (>), less-than (<), and equal (=) symbols:

1 < 3	4 > 2	5 = 5

Your child will know that a whole sandwich is more than a half. The following will introduce your child to the process of comparing fractions — a skill that I cover in Part 4 of this book: "Tackling Third Grade Math"

In this section, you explain to your child how to use these symbols to compare fractions. Worksheet 10-5 at www.dummies.com/go/teachingyourkidsnewmathfd has fractions that your child can compare using the greater-than (>), less-than (<), and equal (=) symbols.

Explain to your child that when you compare fractions, you can use the symbols to show if one fraction is bigger, smaller, or equal to another. Help your child with the fractions shown in Figure 10-8, and then ask them to complete the rest of the problems on Worksheet 10-5.

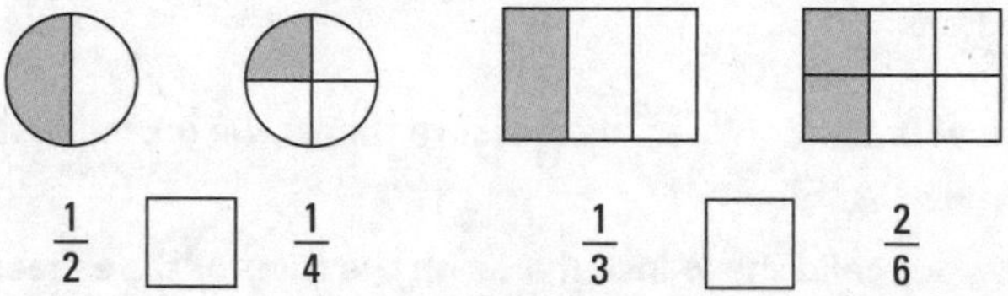

FIGURE 10-8: Comparing fractions using >, <, and =.

$$\frac{1}{2} \quad \square \quad \frac{1}{4} \qquad \frac{1}{3} \quad \square \quad \frac{2}{6}$$

Chapter **11**

Introducing Charts, Graphs, and Word Problems

The amount of data and information in the world is doubling every two years. The vast amount of data facing businesses makes it impossible for pocket-protected engineers to analyze reams of computer printouts to identify business trends. The business-intelligence world has gone visual, and the ability to understand data on charts is a must-have skill.

In this chapter, your child will learn to read simple charts. After that, you will help them with word problems. You remember those: "Two trains left the station at 12:00. . . ." Relax; the word problems in this chapter have no trains!

Before you get started, you'll want to have some blank paper on which you can write some simple charts.

Reading Simple Charts

The ability to read charts and graphs allows your child to apply reasoning skills in a different way than they have with other math concepts. Figure 11-1 shows a simple chart containing the number of books read by three students: Bill, Mary, and Tim.

Look at the figure with your child and follow these steps:

1. **Explain to your child that "a chart shows us information about something. The chart in Figure 11-1, for example, shows us the number of books Bill, Mary, and Tim have read."**

2. **Point to the chart and show your child each student's name.**

   Tell your child that the first bar shows the number of books that Bill read.

3. **Explain to your child that by comparing the height of the bar with the numbers along the line at the left, you can see that Bill read 3 books.**

4. **Ask your child, "How many books did Mary read?"**

   Help your child find Mary's bar on the chart and then use their finger to match the corresponding number.

5. **Repeat this process to identify the number of books that Tim read.**

Figure 11-2 shows a chart that represents how many glasses of lemonade Mary has sold at her lemonade stand over the past three days.

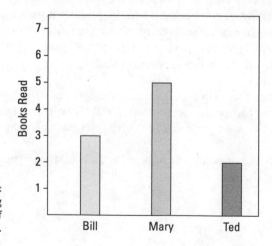

FIGURE 11-1: A chart showing the number of books read.

Point to the chart in Figure 11-2 and show your child the bars that correspond to each day's sales. Then ask your child the following questions:

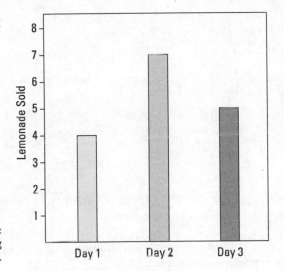

FIGURE 11-2:
A chart showing lemonade sales.

>> On which day did Mary sell the most lemonade?

>> How many glasses did she sell?

>> On which day did Mary sell the least lemonade?

>> How many glasses did she sell?

If your child misses a question, reread the question to them and then help them find the answer by pointing to the correct day and corresponding number of sales.

Figure 11-3 contains a chart that shows the number of apples that Bill, Mary, and Tim picked from a tree.

Introduce the chart in Figure 11-3 to your child and ask them the following questions:

>> Who picked the most apples?

>> How many apples did they pick?

>> Who picked the least number of apples?

>> How many apples did they pick?

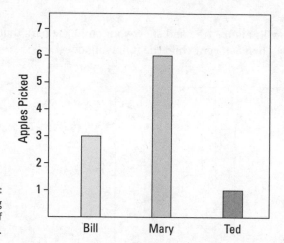

FIGURE 11-3:
A chart showing
the number of
apples picked.

Again, if your child misses a question, reread the question to them and then point to the corresponding information on the chart.

Figure 11-4 contains a different form of chart that shows the number of votes Bill, Mary, and Tim received for class president.

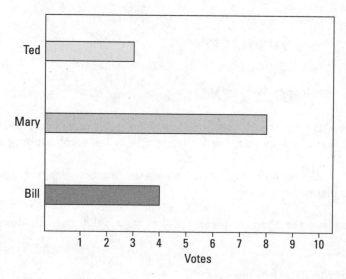

FIGURE 11-4:
A chart showing
votes for class
president.

Introduce the chart in Figure 11-4 to your child. Tell them that each bar shows the number of votes Bill, Mary, and Tim received for class president.

Explain to your child that in this case, the bars are horizontal across the chart but that they can read the chart in the same way as a bar chart. Ask your child the following questions:

>> Who received the most votes?

>> How many votes did they receive?

>> Who received the least number of votes?

>> How many votes did they receive?

Creating Their Own Chart

Figure 11-5 contains a partially complete chart for your child to finish.

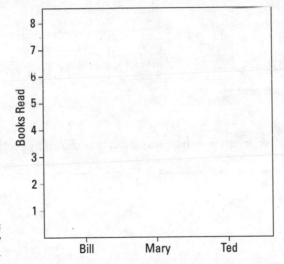

FIGURE 11-5:
A partially
complete chart.

Explain to your child that they are going to complete the chart using the following data:

>> Bill read 5 books.

>> Mary read 7 books.

>> Tim read 3 books.

Use the following steps to help your child work through completing the chart:

1. **Using the chart, have your child find the location for Bill.**

2. **Ask your child to draw a bar for the 5 books Bill read.**

3. **Repeat this process for Mary and Tim.**

   Your chart should look like Figure 11-6.

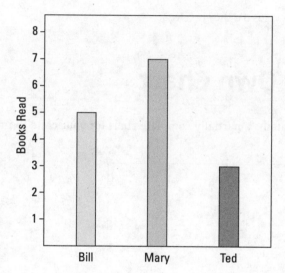

FIGURE 11-6:
A completed chart showing the number of books each person read.

Using a piece of paper, draw a chart similar to that shown in Figure 11-7.

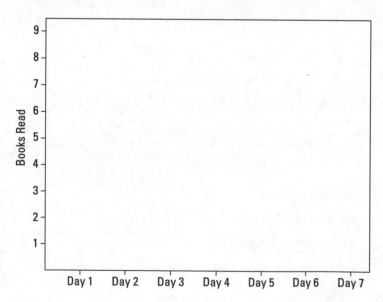

FIGURE 11-7:
The start of a chart to track the number of books read over one week.

Explain to your child that they are going to create a chart that tracks the number of books they read each day for a week.

Show your child the days of the week on the chart and get them to track the number of books they read each day. You might, for example, place a sticky note above each day for each book they have read. Place the chart in a safe location that you can update daily.

Starting with Simple Word Problems

If you are like most people, you may have memories of story problems: "One package leaves New York on a plane while a second package is placed on a bus." Whether those memories are good or bad, word problems are directly applicable to everyday life, and your child will encounter story problems throughout their school years. In this section, you will get started with practicing simple word problems.

Look at the story problem in Figure 11-8 with your child.

FIGURE 11-8:
A simple word problem.

Tim has one dog. Mary has two dogs.

How many dogs do Tim and Mary have?

Using the word problem in Figure 11-8, work through the following steps to come up with a solution:

1. **Have your child read the problem.**

2. **Ask your child to look at the illustration and say, "These are Mary's two dogs and this is Tim's dog. How many dogs do Mary and Tim have altogether?"**

3. **Allow your child to count the dogs.**

4. **Show your child that they can solve the problem using the following expression:**

   $2 + 1 = 3$

Figure 11-9 shows another story problem to examine with your child.

FIGURE 11-9:
A second word
problem.

Bill picked 3 apples, Mary picked 4.

How many apples did Bill and Mary pick?

1. **Have your child read the problem.**

2. **Have your child examine the picture and ask them how they can use addition to solve the problem.**

   You can use flash cards if you'd like.

	+		=	

3. **Help your child as necessary to create the following expression:**

3	+	4	=	7

Figure 11-10 shows one more story problem to practice with your child.

FIGURE 11-10:
A story problem
about the
ever-popular
pizza.

Tim and Bill had 5 slices of pizza. They each ate 1 slice.

How many slices of pizza were left?

Use these steps to help your child determine the answer to the problem:

1. Have your child read the problem.

2. Ask them to examine the picture and tell you how they can use subtraction to solve the problem.

It might help to use flash cards:

	−		=	

3. Help your child as necessary to create the following equation:

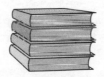

5	−	3	=	2

For the last word problem, shown in Figure 11-11, your child will be able to use the skills for reading and creating charts that were introduced earlier in this chapter.

FIGURE 11-11:
A word problem to be solved with a chart.

Tim read 2 books.   Mary read 4 books.   Bill read 6 books.

Have your child use the chart shown in Figure 11-12 to record how many books each child read and state who read the most books.

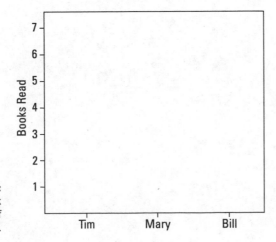

FIGURE 11-12:
A chart to track the number of books read.

3

Advancing with Second Grade Math

Chapter **12**

Starting to Do Math in Their Head

n a world filled with calculators, it's easy to question why kids should be able to do mental math. You may be thinking, "I'll just teach my kid to ask Alexa or Siri!" The reason for learning to solve problems mentally is that kids who do well in math do well in school. There's no long-term research yet regarding the success of those who can question Siri, but we have to assume the results may be different.

By the end of this chapter, your kid is going to start becoming one of "those" kids — you know, the ones who can just do math in their heads. That's pretty cool, if you ask me.

Beginning to Add and Subtract Mentally

This section will increase your child's ability to perform many of these common addition operations in their head. The ability to add and subtract numbers through 20 is key to your child being able to quickly perform operations with larger numbers with regrouping.

Mentally adding numbers through 20

If you completed the first-grade section of this book with your child, the following may be review. Using your 3x5 index cards, create the following flash cards, and then practice the flash cards with your child on a regular basis until they have mastered these addition problems.

Line up your flash cards into rows. The first row has the flash cards 1 + 1, 1 + 2 . . . up to 1 + 10. The next row is the 2 + 1 cards and so on. Here's a small sample:

1 +1	1 +2	1 +3	1 +4	1 +5	1 +6	1 +7	1 +8	1 +9	1 +10	
9 +1	9 +2	9 +3	9 +4	9 +5	9 +6	9 +7	9 +8	9 +9	9 +10	10 +10

TIP

If you write the answers to the problems on the backs of the cards, your child can use the cards to practice on their own and verify their answers.

Adding numbers through 20 using a number line

Your child has learned to use a number line to add small numbers. A number line works for addition problems with larger numbers, too; the number line just has to be longer.

In this section, your child will use a number line to add numbers through 20. They can use the following number line with the numbers 1 through 20:

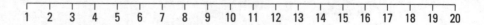

Before you begin the following steps, remind your child that they can use the number line to count, add, and subtract numbers.

1. **Present the following expression to your child:**

9 + 8 =

Remind your child that the first number, 9, tells them where to start on the number line.

2. **Have your child locate the number 9 on the number line.**

3. **Remind your child that the second number, 8, tells them how much to add to 9.**

4. **Have your child use a pencil on the number line to solve the problem.**

   The result should look like this:

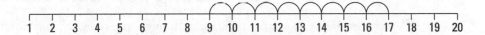

   Download and print Worksheet 12-1 from www.dummies.com/go/ teachingyourkidsnewmathfd to give your child practice with solving addition problems through 20 using the number line. If necessary, help your child complete the first few problems, and then ask them to complete the rest.

Adding numbers through 20 using a deck of playing cards

You've used flash cards for a lot of the exercises in this book, but playing cards are another tool you can use for math practice. It's time to break them out again! Follow these steps for some more addition practice:

1. **Remove the face cards from a deck of playing cards.**

   You can leave the aces and explain to your child that the aces are equal to 1.

2. **Place the cards face down in front of your child.**

3. **Ask your child to turn over two cards and add them.**

4. **If your child is correct, they get to the keep the cards; if they are wrong, use a number line to help them solve the problem, and you keep the cards.**

5. **At the end of the game, the person with the most cards wins.**

Mentally adding numbers through 20 using a worksheet

Your child can use Worksheet 12-2, found at www.dummies.com/go/ teachingyourkidsnewmathfd, a to work on mentally adding numbers through 20. I've included a few problems here that you can help your child solve before you ask them to complete the rest. If your child misses a problem, have them use the number line previously shown in this chapter to solve the problem.

$$3+4$$
$$3+9$$
$$13+6$$

Mentally subtracting numbers through 20

The ability to subtract numbers through 20 is key to your child being able to quickly subtract larger numbers with regrouping (borrowing). If you completed the first-grade section of this book with your child, practicing subtraction should be review.

Pull out all the subtraction flash cards you've created so far (going from $20 - 10$ all the way down to $1 - 1$) or create new cards. Practice with your child to help them become proficient in quickly coming up with the answer to each of the subtraction problems.

TIP

If you write the answer to each problem on the back of the cards, your child can use the cards on their own to check whether they are right. Practice the flash cards with your child on a regular basis until they have mastered these subtraction problems.

Subtracting numbers through 20 using a number line

It's time to revisit subtracting with a number line. Just as your child can add using a number line, they can also use a number line to subtract.

Here's another number line with the numbers 1 through 20:

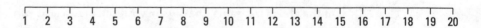

Before you jump into the following exercise, remind your child that they can use the number line to count, add, and subtract numbers. Then you can get started with these steps.

1. **Present the following expression to your child:**

17 - 9 =

2. **Remind your child that the first number, 17, tells them where to start on the number line.**

3. **Have your child locate the number 17 on the number line.**

4. **Then tell your child that the second number, 9, tells them how much to subtract from 17.**

5. **Have your child use a pencil on the number line to solve the problem.**

    The solution should look like this:

17 – 9 =

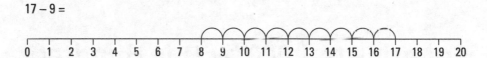

Worksheet 12-3 at www.dummies.com/go/teachingyourkidsnewmathfd includes subtraction problems through 20 that your child can solve with the number line. If necessary, help your child complete the first few problems, and then ask them to complete the rest.

Subtracting numbers through 20 using a deck of playing cards

It's time to make use of the playing cards again — this time, for subtraction. Here's how to play a game to practice subtracting through 20:

1. **Remove the face cards from a deck of playing cards.**

    You can leave the aces and explain to your child that the aces equal 1.

2. **Place the cards face down in front of your child.**

3. **Ask your child to turn over two cards and subtract the smaller number from the larger number.**

4. **If your child is correct, they get to the keep the cards; if they are wrong, again use a number line to help them solve the problem, and you keep the cards.**

5. **At the end of the game, the person with the most cards wins.**

Mentally subtracting numbers through 20 using a worksheet

Worksheet 12-4 at www.dummies.com/go/teachingyourkidsnewmathfd is a tool your child can use to practice mentally subtracting numbers through 20. Here are a few problems you can help your child with if necessary, and then ask them to complete the rest:

$7 - 2$
$10 - 1$
$13 - 6$

Counting to 1,000 by One Hundreds

Your child knows how to count to 100 by twos, fives, and tens. In this section, your child will work up to a really big number — 1,000! — and it's helpful to be able to count by one hundreds to get there.

1. **Create the following flash cards:**

100	200	300	400	500	600	700	800
900	1,000						

2. **Present the cards to your child and tell them that they are going to learn to count to 1,000 by 100s.**

3. **Point to and say each number and then ask your child to do the same.**

4. **Turn the 100-card face down.**

5. **Have your child recite the sequence again, saying 100 from memory.**

6. **Turn the 200-card face down and repeat the process.**

7. **Explain to your child that you are going to say a starting number and they will count to that number to 1,000 by 100s.**

   For example, say "300," and your child should complete counting by hundreds from that point through 1,000. Repeat this process with the numbers 700 and 900.

Reviewing the Tens and Ones Places

In the first-grade section of this book, your child learns to identify the tens and ones places for two-digit numbers. In the following example, 5 is in the tens place and 7 is in the ones place:

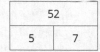

Worksheet 12-5 at www.dummies.com/go/teachingyourkidsnewmathfd gives your child practice with identifying the tens place and the ones place. I've included a few examples here for you to discuss with your child, and then they can complete the worksheet.

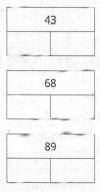

Knowing the Hundreds, Tens, and Ones Places

Just when your child was getting used to the tens and ones places, it's time to throw another into the mix: the hundreds place. The good news is that the technique for learning it is the same, so this may feel like familiar territory to your child.

Before you begin, remind your child that they know how to identify the tens and ones places for a two-digit number, such as 43. Then begin working on the hundreds place with the following steps.

1. **Explain that three-digit numbers have three places: hundreds, tens, and ones.**

2. **Look at the following number with your child and note the hundreds, tens, and ones places:**

427		
4	2	7

Download and print Worksheet 12-6 (www.dummies.com/go/teachingyour kidsnewmathfd) to give your child some practice with identifying the hundreds, tens, and ones places. Help your child complete the first few, and then ask them to complete the worksheet.

538		

764		

123		

Identifying Missing Numbers from 1 to 1,000

You can find Worksheet 12-7 at www.dummies.com/go/teachingyourkids newmathfd, which contains a grid of numbers from 1 through 1,000 that your child can use to fill in missing numbers. Help your child identify the first few missing numbers and then ask them to complete the worksheet. If your child misses a number, have them start counting the numbers from the start of the row to identify the missing number.

Adding and Subtracting Large Numbers without Regrouping

Now that your child is starting to become more comfortable with larger numbers, you can begin working on adding and subtracting them without regrouping. The approach is the same: First, you'll start with using boxes, and then you'll practice working without the boxes to keep track of the hundreds, tens, and ones places.

Adding using boxes

When you and your child worked on adding two-digit numbers, they used boxes to help them track the tens and ones places. In this section, they use boxes again — this time for the hundreds, tens, and ones places. Here's how to work through the process.

1. **Present the following expression to your child and have your child read the numbers out loud:**

	2	3	1
+	4	4	7

2. **Explain to your child that to add large numbers, they start with the ones column:**

	2	3	1
+	4	4	7
			8

3. **They add the tens column:**

	2	3	1
+	4	4	7
		7	8

4. **They add the hundreds column:**

	2	3	1
+	4	4	7
	6	7	8

5. **Repeat this process for the following expression:**

	5	2	1
+	3	5	8

Your child should get:

	5	2	1
+	3	5	8
	8	7	9

Worksheet 12-8 at www.dummies.com/go/teachingyourkidsnewmathfd provides more problems through 1,000 to be solved using boxes. Help your child complete the first few problems, and then ask them to complete the worksheet.

Adding without boxes

In the previous section, your child solved addition problems that helped them write the tens and ones digits in the correct locations. In this section, your child will learn to add three-digit numbers without the need for boxes to help them with the hundreds, tens, and ones places.

1. **Present the following expression to your child and again have your child read the number out loud:**

   357
   +222

   Explain to your child that they will perform the same steps they used for adding with boxes: adding the ones, then the tens, and then the hundreds, but without the boxes this time.

2. **Have your child write their answer.**

   They should get the following solution:

   357
   +222
   579

Worksheet 12-9 at www.dummies.com/go/teachingyourkidsnewmathfd has addition problems for three-digit numbers without regrouping. Help your child solve the first few problems, and then ask them to complete the worksheet.

TIP

Do not continue with subtraction until your child has mastered addition.

Subtracting three-digit numbers using boxes

In this section, your child will learn to subtract three-digit numbers that don't require regrouping (borrowing for subtraction). To help them line up the numbers correctly in the hundreds, tens, and ones columns, they will first use boxes.

Present the following expression to your child and have your child read the numbers out loud:

	5	3	1
−	4	2	0

1. Explain to your child that to subtract large numbers, they start with the ones column just as they did with addition:

	5	3	1
−	4	2	0
			1

2. They subtract the tens column:

	5	3	1
−	4	2	0
		1	1

3. They subtract the hundreds:

	5	3	1
−	4	2	0
	1	1	1

4. Repeat this process for the following expression:

	7	5	7
−	3	3	6

Your child should get:

	7	5	7
−	3	3	6
	4	2	1

Worksheet 12-10 at www.dummies.com/go/teachingyourkidsnewmathfd contains subtraction problems through 1,000 that your child can solve using boxes. Help your child complete the first few problems, and then ask them to complete the worksheet.

Subtracting without boxes

In this section, your child will subtract three-digit numbers without using boxes to help align the hundreds, tens, and ones places.

Present the following expression to your child:

357
−222

1. **Explain that solving this expression works as it did when the numbers were in boxes: they subtract the ones, then the tens, and then the hundreds.**

2. **Have your child write their answer.**

They should get:

357
−222
135

Worksheet 12-11 at www.dummies.com/go/teachingyourkidsnewmathfd is available for your child to use for practicing subtraction of three-digit numbers without regrouping. Help your child solve the first few problems, and then ask them to complete the worksheet.

Revisiting Mental Addition and Subtraction

The ability to perform common math operations mentally is a skill that will benefit your child as they work with larger numbers and perform multiplication and division. This section offers practice for performing common operations mentally.

Mentally adding 10

Adding 10 to a number is one of the easier forms of multi-digit addition because the digit in the ones place doesn't change. Normally, just the tens place changes, unless you're adding 10 to a number in the nineties. Then, of course, you have to get the hundreds digit involved.

Present the following problems to your child:

	7	5
+	1	0

	4	3
+	1	0

	1	7
+	1	0

Have your child solve the problems. They should get the following answers:

	7	5
+	1	0
	8	5

	4	3
+	1	0
	5	3

	1	7
+	1	0
	2	7

Point out to your child that when they add 10 to any number, the ones digit does not change. Only the tens digit increments by 1.

With this knowledge, your child may be able to quickly add the following expressions even without the boxes:

$$\begin{array}{cccc} 23 & 32 & 44 & 57 \\ +10 & +10 & +10 & +10 \end{array}$$

Mentally subtracting 10

It turns out that subtracting 10 from a number is also pretty easy because in most cases, only the tens digit changes.

Present the following problems to your child:

7	5
− 1	0

4	3
− 1	0

1	7
− 1	0

Have your child solve the problems. They should get:

7	5
1	0
6	5

4	3
1	0
3	3

1	7
1	0
0	7

Point out to your child that when they subtract 10 from any number, the ones digit does not change. Only the tens digit decreases by 1. When they realize this, your child may be able to quickly subtract the following:

$$\begin{array}{cccc} 23 & 32 & 44 & 57 \\ -10 & -10 & -10 & -10 \end{array}$$

Mentally adding 100 to a number

If it's starting to feel like déjà vu, that's good! Part of becoming proficient in math is recognizing how to apply knowledge to similar, but not identical, situations. When you add 100 to a number, normally only the hundreds digit changes.

Present the following problems to your child:

7	5	2
+ 1	0	0

4	3	1
+ 1	0	0

1	7	8
+ 1	0	0

Have your child solve the problems. They should come up with the following answers:

	7	5	2
+	1	0	0
	8	5	2

	4	3	1
+	1	0	0
	5	3	1

	1	7	8
+	1	0	0
	2	7	8

Point out to your child that when they add 100 to any number, the tens and ones digits do not change. Only the hundreds digit increments by 1.

With this knowledge, your child may be able to quickly add the following:

$$\begin{array}{cccc} 233 & 327 & 445 & 576 \\ +100 & +100 & +100 & +100 \end{array}$$

Mentally subtracting 100 from a number

Subtracting 100 from a number is like seeing the negative of a photo (not that kids today know anything about film cameras and negatives). When you subtract 100 from a number, the tens and ones digits don't change.

Present the following problems to your child:

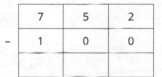

	7	5	2
−	1	0	0

	4	3	1
−	1	0	0

	1	7	8
−	1	0	0

Have your child solve the problems. They should come up with the following answers:

	7	5	2
−	1	0	0
	6	5	2

	4	3	1
−	1	0	0
	3	3	1

	1	7	8
−	1	0	0
	0	7	8

Point out to your child that when they subtract 100 from any number, the tens and ones digits do not change — only the hundreds digit is reduced by 1. If they understand this concept, then your child may be able to quickly subtract the following:

$$\begin{array}{cccc} 233 & 327 & 445 & 576 \\ -100 & -100 & -100 & -100 \end{array}$$

Comparing Large Numbers

A bit of time may have passed since you and your child last talked about comparing numbers, so you may need to remind them of the meanings of the greater-than (>), less-than (<) and equal (=) symbols. After that quick refresher, you can look at the following comparisons:

7 > 3	13 > 11	12 < 21	12 = 12

Have your child use the appropriate symbol to compare the following numbers:

531		278
312		674
412		399

Visit www.dummies.com/go/teachingyourkidsnewmathfd to download and print Worksheet 12-12, which your child can complete to practice comparing large numbers. Help your child with the first few, and then ask them to complete the rest.

Chapter **13**

Having Some Fun with Money and Calendars

I f a penny saved is a penny earned, being able to count them is a good thing. In this chapter, your child will start to make cents, or rather sense, of money. After that, you will teach your child to understand the days of the week and months of the year, as well as how to use a calendar. So, today's a good day to get started, and shortly, your kid will be able to find today on a calendar.

TIP

Before you start, you will want to have the following supplies available:

>> 100 pennies

>> 20 nickels

>> 10 dimes

>> 4 quarters

>> 1 dollar bill

>> A deck of 3x5 index cards

>> A calendar

Counting Change

Money makes the world go 'round — or, at least, that's what some people say. Regardless of the truth of that statement, it's an important life skill for your child to understand units of money.

Explain to your child that when you go to the store and shop, you must pay for the things you buy, and different things cost different amounts of money. Then use these steps to start teaching your child to count money:

1. **Show your child a penny and explain that the coin is a penny and is worth 1 cent.**

   Explain that you count a penny as one. Say, "You can't buy much with 1 penny, but if you have several, you may be able to buy gum or candy."

2. **Place 5 pennies in front of your child and have them count them.**

   Explain to your child that they have 5 cents. Give your child 3 more pennies and ask them how much they now have.

3. **Have your child count the pile of 100 pennies and introduce the dollar bill to your child, saying, "This is a one-dollar bill. It is worth the same as 100 pennies."**

4. **Ask your child to make 10 piles of 10 pennies.**

5. **Point to each pile of 10 pennies and count out loud, "10 cents, 20 cents, 30 cents, . . . one dollar."**

Counting pennies on a worksheet

Worksheet 13-1 at www.dummies.com/go/teachingyourkidsnewmathfd includes illustrations of different numbers of pennies your child can count. Help your child get started, and then ask them to complete the worksheet.

Understanding the value of a nickel

Of course, your child's pockets won't be able to hold enough pennies for them to make purchases, so they need to become familiar with the other coins. Now that your child understands pennies, it's time to move on to nickels with these steps:

1. **Have your child make a pile of 5 pennies.**

   Point to and count the pennies: "1, 2, 3, 4, 5 cents."

2. **Place a nickel near one pile of pennies.**

   Introduce the nickel to your child, saying it is worth 5 cents. In other words, 1 nickel is the same as 5 pennies.

3. **Hand the nickel to your child and tell them that they now have 5 cents.**

4. **Give your child 2 more pennies and ask how much change they have.**

   Point to each coin and help your child count the coins: "5 cents, 6 cents, 7 cents."

5. **Give your child a second nickel and ask them to count their change.**

6. **Repeat this process, giving your child different numbers of nickels and pennies.**

7. **Place 20 nickels in a pile and have your child count them.**

8. **Hold up the dollar bill and tell your child that 20 nickels equals one dollar.**

   Count the nickels out loud: "5, 10, 15, . . . 95, one dollar."

Your child can use Worksheet 13-2, found at www.dummies.com/go/teachingyour kidsnewmathfd, to practice counting combinations of nickels and pennies to determine the correct amount of change. Help your child get started, and then ask them to complete the worksheet.

Knowing that a dime is worth 10 cents

Why do people call the process of providing their opinion, "giving their 2 cents?" Never mind, such questions are a dime a dozen. That said, it's time to move on to dimes with these steps:

1. **Have your child create a pile of 10 pennies.**

2. **Introduce a dime by saying, "This is a dime. It is worth 10 cents."**

3. **Place 10 dimes on the table and count them out loud: "10 cents, 20 cents, 30 cents, . . . one dollar."**

4. **Hold up the dollar bill and tell your child that 10 dimes is equal to one dollar.**

5. **Give your child 1 dime and 2 pennies and explain that they have 12 cents.**

   Point to and count the coins out loud: "10, 11, 12."

6. **Give your child another dime and 2 more pennies.**

   Count the coins out loud with your child: "10, 20, 21, 22, 23, 24."

7. Give your child a nickel and have them count the coins.

8. Repeat this process for different combinations of dimes, nickels, and pennies.

TIP

Counting change takes practice; you will need several sessions for your child to get the hang of it. If your child has difficulty working with several coin types, take a step back and work with pennies, nickels, and dimes one at a time before you try combining them.

Worksheet 13-3, which you can download at www.dummies.com/go/teachingyour kidsnewmathfd, shows different combinations of dimes, nickels, and pennies. Help your child get started with counting the coins on the worksheet, and then ask them to complete the rest of the problems.

Knowing that a quarter equals 25 cents

With pennies, nickels, and dimes under their belt, you can now teach your child about quarters.

1. Have your child create a pile of 25 pennies.

2. Introduce a quarter and say, "This is a quarter. It is worth 25 cents."

3. Place 4 quarters on the table and count them out loud: "25 cents, 50 cents, 75 cents, one dollar."

4. Hold up the dollar bill and tell your child that 4 quarters is equal to one dollar.

5. Give your child a quarter and 4 pennies, counting the coins out loud: "25 cents, 26, 27, 28, 29."

6. Give your child a dime and nickel and have them count the change starting with the quarters, then dimes, nickels, and pennies.

7. Repeat this process for different combinations of quarters, dimes, nickels, and pennies.

Download and print Worksheet 13-4 from www.dummies.com/go/teachingyour kidsnewmathfd for your child to use to practice counting combinations of quarters, dimes, nickels, and pennies. Help your child get started and then ask them to complete the worksheet. If your child misses a problem, use your change to count out the corresponding number of coins with them.

Counting the correct amount of change

Asking your child to count the correct amount of change is a great way to improve not only their knowledge of money but also their overall math skills as well. Follow these steps to practice this skill:

1. **With the coins in front of your child, say, "I'm going to tell you an amount, and you are going to hand me the correct amount of change."**

    Ask your child to give you 4 cents. If they complete the task successfully, move on with the following amounts:

    - 8 cents
    - 12 cents
    - 26 cents

2. **After your child understands the process of counting change, repeat the process with larger amounts.**

3. **If your child does well, place the dollar bill on the table and ask them to add the appropriate change for the following amounts:**

    - $1.07
    - $1.25
    - $1.52

    You can continue practicing amounts until you feel that your child has a firm grasp on combining the coins to get the correct sum of money.

TIP

If you have bills available, you can use the same process to introduce the $5, $10, and $20 bills.

Understanding Calendars

In first grade, your child learned to tell time. A similarly important skill is being able to use a calendar. Your child will understand the number of days in a week and in each month, the number of months in a year, and the number of days in a year. If your child knows these items, you can go to Chapter 14.

Knowing the days of the week

Before you introduce your child to calendars and months, make sure that they know the days of the week. For these steps, you'll break out the old flash cards again.

1. **Use your 3x5 index cards to create flash cards for the days of the week:**

Sunday	Monday	Tuesday	Wednesday	Thursday	Friday	Saturday

2. **Place the cards in order in front of your child, from Sunday through Saturday. Tell your child that each week has seven days.**

3. **Read each card with your child.**

4. **Turn the Sunday card face down and again say the days, having your child say Sunday from memory.**

5. **Repeat this process, turning over all the cards one by one.**

6. **Mix up the cards and have your child place them back in order.**

7. **Have your child count the cards, and tell them that there are 7 days in a week.**

Knowing the months of the year

After your child has mastered the days of the week, you should introduce the months of the year. Tell your child that there are 12 months in a year.

1. **Using your 3x5 index cards, create the following cards:**

January	February	March	April	May	June
July	August	September	October	November	December

2. **Say the months of the year with your child.**

3. **Turn the January card face down and have your child repeat the months, reciting January from memory.**

4. **Repeat this process for each card until your child masters the months.**

5. **Turn all the cards face up, shuffle them around into a random order, and help your child place them back in order.**

TIP

Your child should now know that there are 7 days in a week and 12 months in a year. You should also tell your child that a year has 365 days, unless of course it's a leap year — which we don't discuss until sixth grade, which is not covered in this book!

Understanding a calendar

Explain to your child that you use calendars to know the current date and to track events that will happen in the future. Follow these steps to discuss how days and months appear in a calendar:

1. Show a calendar to your child and then show them a specific month.

2. Say to your child, "The calendar shows the days of the week here. This is the month of ____. We can see from the calendar that it has __ days."

3. Tell your child that each row on the calendar corresponds to one week.

4. Ask your child to count the number of weeks in the current month.

5. Say to your child, "Each box on the calendar corresponds to a specific date. This box, for example, corresponds to (month and day)."

6. Say to your child, "Let's use the calendar to look up your birthday, which is ___. To start, we must turn to the month of ___. Then, we can find the specific day."

   Use the calendar to locate your child's birthday.

7. Say, "Let's find today on the calendar. Today is (month and day)."

   Help your child locate today's date on the calendar.

TIP

A great way to reinforce calendar concepts with your child is to use the calendar regularly to track family events, such as swimming lessons or the days with no school.

Writing a date

Explain to your child that you often need to write down specific dates, such as the day they turned in a school assignment or the date on which you wrote a check. To do so, you first write the month, such as

September

Then, you write the day, followed by a comma and the year:

September 30, 2023

Help your child write the date of their birthday. Finally, help your child write today's date.

Chapter **14**

Revisiting Fractions

I f Arnold Schwarzenegger were an action math figure, he'd be a fraction that says, "I'll be back!" Fractions have returned in this chapter, and they'll be back again in fourth- and fifth-grade math, too! The good news is most of the fraction work in this chapter will be a review (for you as well as your kid). After that, you'll teach your child about *equivalent expressions* — that's a fancy way to say two expressions that have the same value. So, it's time to press Pause on the TV remote and get started.

Before you get started, make sure that you have some scratch paper and a pencil on hand.

Reviewing Fractions

Before your child jumps into new concepts with fractions, it's important that you pump the brakes and review what they have previously learned. Present the following to your child and ask them to identify what fraction of the circle is shaded.

Remind your child that the top number for a fraction specifies the number of shaded parts and the bottom number specifies the total number of parts. In this case, the fraction is

$$\frac{1}{2}$$

Here is another shape with shaded parts:

The fraction that describes this shape is

$$\frac{3}{4}$$

Worksheet 14-1 at www.dummies.com/go/teachingyourkidsnewmathfd includes a collection of shapes your child can work through to identify fractions. Work with your child to complete the worksheet.

Mixing Things up with Mixed Numbers

Remember, a mixed number contains a whole part and a fractional part. Here are some examples:

$$1\frac{1}{4} \quad 3\frac{1}{3} \quad 2\frac{1}{2}$$

Show your child Figure 14-1 to help them visualize these fractions.

FIGURE 14-1:
Shaded whole
shapes and
pieces to
illustrate mixed
numbers.

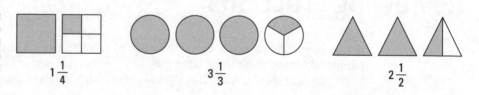

Worksheet 14-2 at www.dummies.com/go/teachingyourkidsnewmathfd includes more visualizations like Figure 14-1. Help your child with the first few problems, and then ask them to complete the rest of the worksheet.

Understanding What Equivalent Fractions Are

Present the fractions in Figure 14-2 to your child and say, "You have learned that fractions that look different can actually be equal." Explain to your child that as they perform arithmetic operations such as addition and subtraction with fractions, they may get results such as $\frac{3}{6}$, $\frac{2}{4}$, or $\frac{1}{2}$, which, if we drew them, would all look the same — making them equivalent fractions.

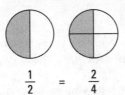

FIGURE 14-2:
The numbers are different, but the shaded area is the same.

$$\frac{1}{2} = \frac{2}{4}$$

Download and print Worksheet 14-3 from www.dummies.com/go/teachingyour kidsnewmathfd. On this worksheet, your child can circle the shape pairs that are equivalent fractions. Help your child get started, and then ask them to complete the worksheet.

Comparing Fractions

Two fractions can be equal, but that's not always the case. Sometimes, one fraction is smaller or larger than the one it's being compared to. Remind your child of the greater-than (>), less-than (<), and equal (=) symbols. You can show them the example in Figure 14-3.

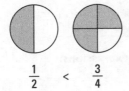

FIGURE 14-3:
Comparing two fractions.

$$\frac{1}{2} < \frac{3}{4}$$

You can use Worksheet 14-4 at www.dummies.com/go/teachingyourkidsnew mathfd for comparing fractions. Help your child get started, and then ask them to complete the worksheet.

Creating Equivalent Expressions

Equivalent expressions. . . . . Sometimes the terms are harder to remember than the concepts themselves. Remind your child that they have learned that different fractions, such as $\frac{3}{6}$, $\frac{4}{8}$, and $\frac{1}{2}$ are equivalent fractions — which means that they are the same. In a similar way, two different expressions, such as 3 + 5 and 4 + 4, can have the same result, in this case 8, so they're equivalent expressions.

Creating equivalent expressions using addition

In this section, your child will learn how to create two expressions that are the same, meaning they're equivalent expressions. Follow these steps:

1. **Use your 3x5 index cards to create the following cards, or pull the following numbers and symbols from your stack of previously used cards:**

0	1	2	3	4	5	6	7	8	9	10
+	+	–	–	=						

2. **Place the cards for the following expression in front of your child:**

3	+	5	=	2	+	

3. **Have your child add 3 + 5 and write down the number 8 on a piece of scrap paper.**

4. **Ask your child, "What number plus 2 equals 8?" If your child says 6, have them place the 6 card down to complete the expression. If they are wrong, have them subtract 8 – 2.**

   They should get the following solution:

3	+	5	=	2	+	6

5. **Repeat this process for the following expressions:**

3	+	4	=	5	+	

6	+	2	=	3	+	

9	+		=	6	+	4

For each expression, have your child start by solving the part of the expression that is complete. Then they can determine which value to add to the given number that will equal the previous result.

I've included a few equivalent expressions problems from Worksheet 14-5, found at www.dummies.com/go/teachingyourkidsnewmathfd, for you to practice with your child. After you've helped them with the following, ask them to complete the worksheet:

3	+	4	=	5	+	

2	+	1	=	1	+	

13	+	4	=	5	+	

Creating equivalent expressions using subtraction

Like with addition, you can create equivalent expressions with subtraction. Using the following steps, you will bring subtraction into the mix:

1. **Using cards you've previously made, place the following expression in front of your child:**

5	–	2	=	4	–	

2. **Have your child solve 5 – 2 and write down the number 3 on a piece of scrap paper.**

3. **Ask your child, "4 minus what number equals 3?"**

   They should get the following solution:

5	–	2	=	4	–	1

4. **Repeat this process for the following expressions:**

7	–	4	=	5	–	

9	–	2	=	8	–	

5	–		=	6	–	4

For each expression, have your child first solve the part of the expression that is complete. Then they can determine what value subtracted from the given number in the incomplete expression will result in a number equal to the previous result.

I've included a few equivalent expressions from Worksheet 14-6, located at www.dummies.com/go/teachingyourkidsnewmathfd, for you to practice with your child. After you've helped them with the following, ask them to complete the worksheet:

8	+	4	=	5	+	

2	+	1	=	1	+	

13	+	4	=	5	+	

Creating Equivalent Expressions Using Addition and Subtraction

Up to this point, your child has done addition and subtraction separately, so they may be amazed to find out that they can create equivalent expressions using both addition and subtraction.

1. **Using cards you've previously made, place the following expression in front of your child, noting that the expression contains both addition and subtraction:**

3	+	5	=	10	−	

2. **To start, have your child add 3 + 5 and write down the number 8 on their scrap paper.**

3. **Ask your child, "10 minus what number is 8?"**

They should get the following answer:

3	+	5	=	10	−	2

4. Repeat this process for the following expressions:

3	+	4	=	10	–	

6	+	1	=	9	–	

9	+		=	10	–	0

5. For each expression, have your child first solve the part of the expression that is complete. Then they can determine what value added to the given number will equal the previous result.

I've included a few equivalent expressions problems from Worksheet 14-7 from www.dummies.com/go/teachingyourkidsnewmathfd for you to practice with your child. After you've helped them with the following, ask them to complete the worksheet:

8	+	4	=	15	–	

11	+	4	=	19	–	

16	–	14	=	13	–	

IN THIS CHAPTER. . .

» Reviewing adding and subtracting with carrying and borrowing

» Learning old school carrying and borrowing

» Self-checking answers

» Adding and subtracting large numbers using a number line

Chapter **15**

Adding and Subtracting with Regrouping

Regrouping — apparently, it's become a term for the ages. Back in the day, you learned to add and subtract using carrying and borrowing. Before that, Einstein and Newton used carrying and borrowing. But it's a new world, and it turns out (as you'll find out in this chapter) there are new-math techniques that let kids add and subtract without regrouping. That said, old habits die slowly, and this chapter starts with a few old-school techniques. After that, I show you the new-school techniques, which are actually pretty cool.

TIP

Before you get started, you'll want to have the following supplies on hand:

>> Sheets of paper and a pencil

>> Straws and rubber bands

Reviewing Addition without Regrouping

Before you jump into regrouping, let's do a review of two-digit addition that does not require it. Here's an example you can work through with your child as a refresher:

$$\begin{array}{r} 23 \\ + 32 \\ \hline \end{array}$$

Remind your child that to add two-digit numbers, they first add the ones column and then the tens column. When you ask your child to solve the problem, they should get this answer:

$$\begin{array}{r} 23 \\ + 32 \\ \hline 55 \end{array}$$

Worksheet 15-1 at www.dummies.com/go/teachingyourkidsnewmathfd is good practice for two-digit addition without regrouping. I've included the first three problems here. Help your child with these, and then ask them to complete the rest of the worksheet.

$$\begin{array}{ccc} 25 & 42 & 33 \\ +31 & +41 & +66 \\ \hline \end{array}$$

Using Boxes to Solve Addition Problems that Require Regrouping

This section introduces regrouping (carrying) to your child. Learning to add and subtract with regrouping is a challenging process, so to help your child with the carrying process, I've included boxes, within which they can write the value they are carrying. I've also included practice problems throughout this chapter for you to work through with your child.

Present the following problem to your child and then work through the steps:

$$\begin{array}{r} \square\ \square \\ 4\ \ 7 \\ +\ 3\ \ 5 \\ \hline \end{array}$$

1. Say to your child, "To solve this problem, start with the ones digits. In this case, you will add 7 + 5, which is 12."

2. Explain to your child, "When the result of your addition is larger than 9 (as 12 is), you write the 2 in the ones place and carry the 10 to the tens column by writing a 1 within the box:"

```
   □  1
   4  7
 + 3  5
      2
```

3. Have your child solve the tens column.

Explain that if they've put a 1 in the box, they add it to the other digits in that column. In this case, 1 + 4 + 3 is 8, which your child should write:

```
   □  1
   4  7
 + 3  5
   8  2
```

Present the following expression and then work through the steps for more practice of addition with regrouping:

```
   □  □
   3  6
 + 4  8
```

1. Remind your child to start in the ones column to add the numbers 6 + 8.

The sum is 14, so your child should write the 4 in the ones column and write the 1 in the box above the tens column, like so:

```
   □  1
   3  6
 + 4  8
      4
```

2. To complete the problem, your child should add the tens digits 1 + 3 + 4, which equal 8, as shown here:

```
   □  1
   3  6
 + 4  8
   8  4
```

Present the following expression:

```
    ☐ ☐
      3 7
    + 3 3
```

Have your child solve the problem. They should come up with the following solution:

```
    ☐ 1
      3 7
    + 3 3
      7 0
```

Regrouping applies to more than problems with two-digit solutions. Sometimes, your child will have to regroup to both the tens and the hundreds places. Present the following problem:

```
    ☐ ☐
      6 3
    + 3 9
```

1. **Remind your child to start with the ones place.**

    Have your child add 3 + 9, which is 12. Have them write the 2 in the ones column and carry the 1:

    ```
        ☐ 1
          6 3
        + 3 9
            2
    ```

2. **Remind your child to add the digits in the tens column.**

    Your child should add 1 + 6 + 3, which is 10. In this case, they will write the 0 and carry 1 to the hundreds column, as shown here:

    ```
        1 1
          6 3
        + 3 9
          0 2
    ```

3. Say to your child, "In this case, you carried from the tens digits to the hundreds digits. Because you don't have any other digits in that column to add, you simply bring down the 1."

```
  ⬚ ⬚
    6 3
  + 3 9
  ───────
  1 0 2
```

See if your child can work through another problem of regrouping of the tens and hundreds places by presenting the following expression:

```
  ⬚ ⬚
    7 7
  + 8 5
```

They should come up with the following solution:

```
  ⬚ ⬚
    7 7
  + 8 5
  ───────
  1 6 2
```

Download and print Worksheet 15-2 from Dummies.com for your child to use to practice adding with boxes and carrying. Help your child with the first few problems, and then ask them to complete the worksheet.

Solving Addition Problems that Require Regrouping without Boxes

In the previous section, you used boxes to help your child know where to write the value that they carry. In this section, you show your child how to work without the boxes. You may need to help your child to align the values over the tens place.

Explain to your child that they have learned to add multi-digit numbers by placing the carry digit in a box above the tens and hundreds units and that the next step is to add without the boxes. You can remind them of earlier lessons when they first added with boxes and then practiced adding without the boxes. Then work through these steps to show them the method:

1. **Tell your child that when they must carry a number, they will simply write the carry digit above the number in the location where the box would have been.**

   Here's an example problem to work with:

   ```
     57
   + 33
   ```

2. **Explain to your child that when they add 7 + 3, they get 10.**

   They need to write the 0 in the ones column and carry the 1 by writing it above the tens column, as shown here:

   ```
     1
     57
   + 33
     90
   ```

3. **Present the following problem to your child:**

   ```
     56
   + 37
   ```

4. **Ask your child add the ones column. They should get 13. Have your child write the 3 beneath the underline and carry the 1 by writing it above the 5:**

   ```
     1
     56
   + 37
      3
   ```

5. **Have your child add the tens column.**

   They should get the following result:

   ```
     1
     56
   + 37
     93
   ```

Adding Numbers Using Decomposition

Decomposition is the process of breaking a number into hundreds, tens, and ones digits. In this section, your child will learn how to decompose a number and then use the parts to simplify addition. Use these steps to work through this process.

1. **Say to your child, "Consider the following number."**

    87

2. **Explain to your child that they can decompose the number 87 by breaking it into its tens and ones parts, like so:**

    $87 = 80 + 7$

3. **Present the following number to your child and ask them to decompose it:**

    357

    They should break it down like this:

    $357 = 300 + 50 + 7$

Addition problems can be worked by decomposing the numbers in the expression. Here's an example to work through with your child:

1. **Present the following expression to your child:**

    35
    + 57

2. **Ask your child to decompose each of the numbers in the expression:**

    35    $30 + 5$
    + 57 $50 + 7$

3. **Ask your child to add the decomposed numbers:**

    $30 + 5$
    $50 + 7$
    $80 + 12 = 92$

4. **Ask your child to solve the following problem using decomposition:**

    352
    + 579

They should get the following result:

$$300 + 50 + 2$$
$$+500 + 70 + 9$$
$$800 + 120 + 11 = 931$$

To practice addition with decomposition, your child can work through Worksheet 15-3 from www.dummies.com/go/teachingyourkidsnewmathfd. Here's one more problem you can help your child to solve; then ask them to try the rest on their own.

$$\begin{array}{r} 25 \\ + 31 \\ \hline \end{array}$$

 Being able to add multi-digit numbers with regrouping is a key math skill. Do not move on to subtraction until your child has mastered it. Practice the worksheet with your child daily. It may take a week or more of practice until they are ready to move on.

TIP

Reviewing Subtraction without Regrouping

Once your child is comfortable with the idea of adding with regrouping, it's time to work your way into subtraction. However, before you and your child take the plunge into performing subtraction that requires borrowing, you should ease into the math waters by reviewing the process of subtracting the ones and tens digits.

Remind your child that to subtract two-digit numbers, they first subtract the ones column and then the tens column. Use the following problems as a refresher:

$$\begin{array}{cccc} 23 & 29 & 33 & 49 \\ -11 & -14 & -22 & -27 \\ \hline \end{array}$$

Here are the solutions:

$$\begin{array}{cccc} 23 & 29 & 33 & 49 \\ -11 & -14 & -22 & -27 \\ \hline 12 & 15 & 11 & 22 \end{array}$$

Understanding the Concept of "Borrowing"

You've laid the groundwork and done a little warmup. Now, your child is ready to learn how to subtract numbers that require them to borrow from the tens digits. Visual aids are useful for explaining the concept of borrowing, so you will start with bundles of straws:

1. **Use your straws and some rubber bands to create 5 bundles of 10 straws.**

Hand the bundles and 2 individual straws to your child.

2. **Present the following problem to your child:**

$$\begin{array}{r} 52 \\ -\ 37 \\ \hline \end{array}$$

3. **Say to your child, "You have 52 straws. The problem says you must subtract 37 straws and give them back to me. However, you only have 2 individual straws and you must subtract 7. For you to be able to subtract the 7 straws, you will need to borrow and unwrap one of the bundles of 10 straws."**

Hand the unwrapped bundle of 10 straws to your child and ask them to subtract the 7 straws from their collection of 12.

4. **Ask your child, "How many straws do you have left?"**

They should respond by telling you that they have 15 straws left.

5. **Offer your child this explanation: "When we can't subtract one number from another because the top number is too small, we can borrow 10 ones from the tens unit."**

Examine the subtraction problem again and say, "In this case, we can't subtract 7 from 2, so we need to borrow 10 from our tens unit. When we do that, we write a small 1 in front of the 2, making it a 12. Also, because we borrowed 10 straws, we must reduce our tens digit from 5 to 4."

$$\begin{array}{r} \boxed{4} \\ \cancel{5}\,^{1}2 \\ -\ 3\ 7 \\ \hline 5 \end{array}$$

6. Say, "Now you can subtract the tens digits."

$$
\begin{array}{r}
\boxed{4} \\
\not{5}\,{}^1 2 \\
-3\ \ 7 \\
\hline
1\ \ 5
\end{array}
$$

7. Repeat the process of using bundled and single straws for the following expression:

$$
\begin{array}{r}
62 \\
-\ 49 \\
\hline
\end{array}
$$

8. Point out that, as before, you can't subtract 9 from 2, and you must borrow, which creates the following:

$$
\begin{array}{r}
\boxed{5} \\
\not{6}\,{}^1 2 \\
-4\ \ 9 \\
\hline
\end{array}
$$

9. Subtract 12 – 9, writing the answer (3), as shown here:

$$
\begin{array}{r}
\boxed{5} \\
\not{6}\,{}^1 2 \\
-4\ \ 9 \\
\hline
3
\end{array}
$$

10. Finish by subtracting the tens digits:

$$
\begin{array}{r}
\boxed{5} \\
\not{6}\,{}^1 2 \\
-4\ \ 9 \\
\hline
1\ \ 3
\end{array}
$$

Now that your child is familiar with the concept of borrowing, have them solve the following problem:

$$
\begin{array}{r}
27 \\
-\ 19 \\
\hline
\end{array}
$$

They should get this:

```
  1
  2 ¹7
 -1  9
 ───────
  0  8
```

Use Worksheet 15-4 from www.dummies.com/go/teachingyourkidsnewmathfd to practice two-digit subtraction with borrowing. Help your child with the first few, and then ask them to complete the worksheet.

Learning to Check Their Work

Even people who are highly proficient with math can sometimes make mistakes, so it's good to know how to check one's work to see whether the answer is correct. Here's how to introduce this idea to your child and help them get into the habit of checking their own work:

1. **Explain that addition and subtraction are opposites.**

   In other words, if you subtract a number and get a result, you can add the answer with one of the values in the problem to ensure that the result is the other starting number. When it is, you've solved the problem correctly; otherwise, you need to recheck the subtraction.

2. **Present the following problem to your child:**

   ```
     45
   + 22
   ─────
     67
   ```

3. **Tell your child they can make sure that their addition operation is correct by subtracting one of the two numbers they added to see if they get the other number, like so:**

   ```
     67
   - 22
   ─────
     45
   ```

   You can say, "In this case, because our numbers are the same, we know our original addition is correct."

4. **Repeat the process with the following expression:**

$$
\begin{array}{r}
49 \\
-\ 23 \\
\hline
26
\end{array}
$$

Remind your child that they test whether the subtraction is correct by using addition. In this case, they can add the two small numbers to see if they get the larger one:

$$
\begin{array}{r}
26 \\
+\ 23 \\
\hline
49
\end{array}
$$

Again, because the numbers are the same, the previous subtraction is correct.

Using New Math to Add and Subtract Large Numbers

The process of carrying numbers to add and borrowing numbers to subtract is considered passé, dated, and old school. It's just not something most kids learn at school anymore.

Therefore, your child is probably going to bring home math problems for which the approach to solve them seems foreign to you. The following sections will bring you up to speed on several "new-school" approaches to addition and subtraction.

Using an open number line to add numbers

The good news is that the first approach to adding two-digit numbers uses a number line, which should feel familiar. Consider the following problem:

$$
\begin{array}{r}
55 \\
+\ 32 \\
\hline
\end{array}
$$

Using a piece of paper and a pencil, draw a number line and mark the number 55 on it, as shown here:

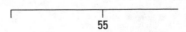

55

Then, draw three loops to indicate the 3 tens in 32. Finally, draw the 2 ones, like this:

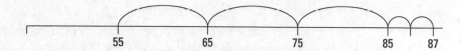

Next, try the process with the following expression:

37
+ 25

Again, draw a number line and mark the number 37. Then, draw the 2 tens and the five ones:

You can also use the number line method with three-digit addition problems. Present the following problem to your child:

347
+ 255

Again, draw a number line and mark the number 347. Then, draw 2 large hundreds loops, 5 loops for the tens, and 5 small ones loops, as shown here:

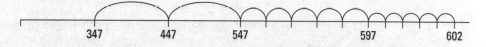

Worksheet 15-5 is set up with addition expressions and open number lines for solving the problems. After working through the first couple of problems with your kiddo, ask them to complete the worksheet. You can download the worksheet from www.dummies.com/go/teachingyourkidsnewmathfd.

Performing subtraction using a number line

Repeatedly throughout this book, you've been able to show your child that the methods for addition work similarly for subtraction. This chapter is no different. Here, you show your child how to use a number line with subtraction expressions.

Present the following expression:

$$
\begin{array}{r}
55 \\
-\ 37 \\
\hline
\end{array}
$$

Draw a number line on a piece of paper and mark the number 55. Then, moving left toward 0, draw loops to subtract the 3 tens, like this:

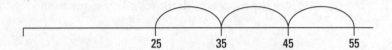

Then, draw seven small loops to represent the 7 ones:

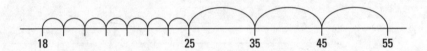

Repeat this process for the following expression:

$$
\begin{array}{r}
81 \\
-\ 44 \\
\hline
\end{array}
$$

You should get:

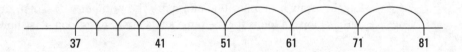

Number lines apply to three-digit subtraction, too. Present the following expression to your child:

$$
\begin{array}{r}
346 \\
-\ 157 \\
\hline
\end{array}
$$

Again, draw a number line and mark the value 346:

Draw a loop subtracting the 100:

Then, draw loops subtracting the 5 tens and 7 ones:

The last worksheet for this chapter, Worksheet 15-6, contains subtraction problems your child can solve using number lines. Help your child with the first few, and then ask them to complete the worksheet. (You can download the worksheet from www.dummies.com/go/teachingyourkidsnewmathfd.)

Checking Their Work Using a Number Line

When your child uses a number line to add numbers, they can check their work using subtraction on a number line. Consider the following expression and ask your child to solve it with a number line:

$$\begin{array}{r} 55 \\ + 32 \end{array}$$

Using the number line, they should get:

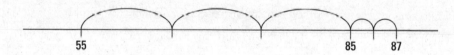

To check the addition work, your child can subtract the top number from their result using a number line. Show your child the following number line, which checks the work of the preceding expression:

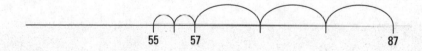

Here's another expression for practice:

$$\begin{array}{r} 63 \\ - 44 \end{array}$$

Have your child use a number line to solve the problem:

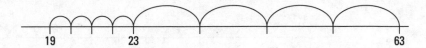

Ask your child to use the number line to add the result back to the bottom number. If their math is correct, they will get the top number, as shown here:

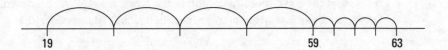

4

Tackling Third Grade Math

Chapter **16**

Introducing Basic Multiplication and Division

I f you thought learning how to multiply and divide was hard, wait until you try to explain it to your kid — oh, wait, that's the purpose of this chapter! But I have good news! Your set of straws will help you to make sense of multiplication and division without much fuss. Multiplication and division are advanced skills, but this chapter breaks down the process, allowing you to divide and conquer!

TIP

Before you get started, you'll want to have the following supplies on hand:

» 100 straws

» Sheets of paper and a pencil

Understanding the Multiplication Process

Now that your child has mastered addition and subtraction, multiplication is next in line. You can think of multiplication as repeated addition.

Tell your child that they are going to learn to multiply numbers.

1. **Introduce your child to the multiplication sign.**

   You might say, "Just as addition operations use the plus (+) sign and subtraction problems use the minus (–) sign, multiplication problems use what we call a times sign, which looks like an x." Show your child this example:

   $1 \times 3 =$

2. **Tell your child that when you multiply numbers, you're working with groups of numbers.**

3. **To illustrate the concept of multiplication, take three straws and place them in front of your child.**

   Tell your child that they have one group of 3 straws. In other words, they can think of 1×3 as meaning "1 group of 3."

4. **Place the following expression in front of your child:**

   $2 \times 3 =$

5. **Hand your child three more straws.**

   Have your child place the three straws they already had into one pile and three new straws in a second pile.

6. **Explain to your child that they have 2 piles of 3 straws.**

   Have your child count the straws to confirm that the two piles contain 6 straws.

7. **Tell your child that two groups of straws represent 2 × 3 = 6.**

8. **Place the following expression in front of your child:**

   $3 \times 3 =$

9. **Hand your child another 3 straws and tell them that they have 3 piles of 3 straws, or 9 straws in all.**

   Tell your child that $3 \times 3 = 9$.

Here's another expression you can solve with groups of straws:

$2 \times 5 =$

Have your child create 2 piles of 5 straws and then ask, "What is 2×5?" Allow your child to count the straws.

Repeat this process for the following equations:

$4 \times 3 =$

$8 \times 1 =$

Using tables to multiply numbers

Using straws to multiply numbers is a great way to learn. Unfortunately, your child won't always have groups of straws in their pocket, so you need to introduce another way to solve multiplication problems. You can show them how to draw a table that helps them solve these problems.

1. **Place the following expression in front of your child:**

   $3 \times 4 =$

2. **Ask your child to read the expression and tell you how many groups of 4 you're building.**

3. **On a piece of paper, draw the following table:**

   $3 \times 4 =$

4. **Explain to your child that rows go across the page and columns go up and down. Have your child count the rows (3) and the number of columns (4).**

5. **To get the solution for 3 × 4, have your child count the number of boxes (12) in the table.**

6. **Tell your child that 3 × 4 = 12.**

7. **Show your child the following expression:**

   $4 \times 5 =$

8. **Ask your child to read the expression and tell you how many bundles of 5 you are building.**

9. Say, "To represent those bundles, we can use this table."

$4 \times 5 =$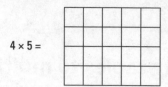

10. Have your child count the number of rows (4) and the number of columns in each row (5).

11. Have your child count the total number of boxes (20) and say, "4 × 5 = 20."

12. For this last expression, ask your child to draw the corresponding table and solve the expression:

$3 \times 3 =$

Visit www.dummies.com/go/teachingyourkidsnewmathfd to download and print Worksheet 16-1. Your child can solve the multiplication problems in this worksheet by drawing the corresponding tables.

Multiplication is repeated addition

Earlier in this chapter, I say that you can think of multiplication as repeated addition. This section helps reinforce that concept:

1. Present the following expression to your child:

$3 \times 2 =$

2. Explain to your child that the first number (3) tells them how many times they are going to add the second number (2):

$3 \times 2 = 2 + 2 + 2 = 6$

3. Present and explain the following expressions to your child:

$4 \times 3 = 3 + 3 + 3 + 3 = 12$
$5 \times 2 = 2 + 2 + 2 + 2 + 2 = 10$
$2 \times 5 = 5 + 5$
$3 \times 1 = 1 + 1 + 1$

4. **Have your child write the corresponding addition expressions for the following multiplication expressions:**

$3 \times 4 =$

$2 \times 3 =$

$4 \times 2 =$

$7 \times 1 =$

They should get the following results:

$3 \times 4 = 4 + 4 + 4 = 12$

$2 \times 3 = 3 + 3 = 6$

$4 \times 2 = 2 + 2 + 2 + 2 = 8$

$7 \times 1 = 1 + 1 + 1 + 1 + 1 + 1 + 1 = 7$

Multiplying by 1

Multiplication doesn't get any simpler than multiplying by 1. After all, 1 times any number is that number.

You've shown your child that multiplying numbers is like creating groups that each have the same number of items in them. Use these steps to explain the idea of multiplying by 1.

1. **Show your child the following expressions:**

$1 \times 1 = \qquad 1 \times 2 = \qquad 1 \times 100 =$

2. **Tell your child that these expressions are just like the ones you saw earlier: the first digit is the number of straws in each group and the second digit is the number of piles.**

If you create 1 pile with 1 straw, you have 1 straw. Likewise, if you create 1 pile of 2 straws, you have 2 straws. Finally, if you create 1 pile of 100 straws, you have 100 straws.

3. **Say, "When we multiply 1 times any number, the result is that number."**

Multiplying by 0

Okay, multiplying 1 times any number is pretty easy! So is multiplying a number times 0 — the result is always 0. In this section, you will teach your child just that.

1. **Show your child the following expressions:**

$$0 \times 1 = \qquad 0 \times 2 = \qquad 0 \times 100 =$$

2. **Ask your child, "How many piles are you creating?"**

   If they answer, "Zero," say, "That's right. If we have 0 piles, we have 0 straws, regardless of the number we are multiplying. That means 0 times any number is 0."

Practicing multiplication flash cards

The ability to multiply the numbers 1 through 10 is key to your child's ability to multiply multi-digit numbers, so it's important that they develop the skill for recognizing problems and quickly knowing the answers.

1. **Using your 3 × 5 index cards, create multiplication flash cards going from 0 × 0 through 10 × 10.**

TIP

   To make it easier for your child to use the flash cards to practice without your assistance, write the answers on the back of each card.

2. **Place the following cards in front of your child and say, "What have you learned about multiplying any number times 0?"**

0 ×0	0 ×1	0 ×2	0 ×3	0 ×4	0 ×5	0 ×6	0 ×7	0 ×8	0 ×9	0 ×10

   If they answer correctly, say, "That's right! Zero times any number is 0. So, all of these cards will equal 0."

   If your child answers incorrectly, revisit the section on multiplying a number by 0.

3. **Place the following cards in front of your child and say, "What have your learned about multiplying any number times 1?"**

1 ×0	1 ×1	1 ×2	1 ×3	1 ×4	1 ×5	1 ×6	1 ×7	1 ×8	1 ×9	1 ×10

4. **Place the following cards in front of your child:**

2 ×0	2 ×1	2 ×2	2 ×3	2 ×4	2 ×5	2 ×6	2 ×7	2 ×8	2 ×9	2 ×10

5. **Remind your child that they already know the answer to the first two cards because they know what it means to multiply a number by 0 or 1.**

   Then say, "When you multiply a number by 2, you double the number. 2 × 2 is 4, 2 × 3 is 6, and so on."

6. **Give the 3 sets of cards to your child to practice with.**

   Remind them that they can always draw boxes to confirm their results, as with this example:

   After your child has mastered the first three sets of cards, give them the fourth set:

3 ×0	3 ×1	3 ×2	3 ×3	3 ×4	3 ×5	3 ×6	3 ×7	3 ×8	3 ×9	3 ×10

Repeat this process of adding on sets of flash cards one at a time until your child has mastered all the flash cards.

TIP

Be patient. It may take your child several lessons before they master the multiplication flash cards. Have your child practice the cards for the numbers they're currently working on each day.

Completing multiplication problems on a worksheet

Worksheet 16-2, found at www.dummies.com/go/teachingyourkidsnewmathfd, includes multiplication problems for your child to solve. If your child has mastered the multiplication flash cards, they should find success with this worksheet.

TIP

On this book's companion website is Worksheet 16-3, another multiplication worksheet. Download and print several copies of the worksheet for your child to use for practice. Time your child as they work. Ideally, they will complete the worksheet in five minutes or less.

Multiplying any number by 10

Your child has learned that multiplying 0 or 1 times any number is easy. It turns out, they can apply the knowledge from those lessons to multiplying a number times 10. Look at the following expressions with your child:

$1 \times 10 =$ $2 \times 10 =$ $3 \times 10 =$

Have your child draw the tables for each operation. Their tables should look like this:

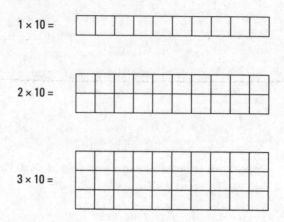

Use the following steps to discuss multiplying by 10 and other numbers that end in 0:

1. Ask your child to examine the results in the tables they've drawn, and then say, "When you multiply any number times 10, you simply add a 0 to the number. For example, 1 × 10 = 10 (add a 0 to 1 to get 10), 2 × 10 = 20, and 3 × 10 = 30."

2. Explain that you can use the process of adding a zero to any number you multiply times 10.

   Here are some examples you can look at:

   $45 \times 10 = 450$
   $37 \times 10 = 370$
   $100 \times 10 = 1,000$

Understanding the Division Process

Now that your child has learned to multiply, the next step is for them to learn division.

Explain to your child that division is the opposite of multiplication. In other words, you are given the total number and number of groups. The goal is to determine the number of items in each group. Here's an example to work through with your child.

1. Give your child 6 straws and show them the following equation:

   $6 \div 2 =$

2. Explain that your child is starting with 6 total straws, and they must divide them into 2 groups.

3. Have your child place 1 straw at a time into 2 groups.

4. When your child has put all 6 straws into the 2 groups, have them count the number of straws in a group.

   In this case, they should have 2 groups of 3 straws each.

5. Tell your child that "6 ÷ 2 = 3."

6. Repeat this process for the following expression:

   $9 \div 3 =$

7. Ask your child, "How many straws will you start with?"

   They should answer, "9."

8. Then ask, "Into how many groups will you divide the straws?"

   They should answer, "3."

9. Have your child place straws one at a time into each of the 3 piles until they are out of straws.

10. When your child has put all 9 straws into the 3 groups, have them count the number of straws in 1 pile.

    They only need to count the straws in one pile because they evenly distributed the straws so that each pile contains the same amount. In this case, they should have 3 piles of 3 straws each.

11. Say, "9 ÷ 3 = 3."

Solving division problems using multiplication

Now that your child has learned how to multiply numbers, it may help them to master division by using multiplication as they divide. With this method, you ask your child to consider what number they would multiply by to get the result. If they're well-versed in multiplication, it may be easier to come up with the answer than if they try to divide two numbers mentally. Here's an example.

1. **Present the following expression to your child:**

   $12 \div 4 =$

2. **Remind your child that they have learned that division is the opposite of multiplication.**

3. **Say, "When you see an expression such as 12 ÷ 4, ask yourself, 'What number times 4 equals 12?'"**

   $3 \times 4 = 12$

4. **Present the following problem to your child:**

   $20 \div 4 =$

5. **Ask, "What number times 4 equals 20?"**

   If your child answers "5," then you can say, "Correct. You know that $4 \times 5 = 20$ so $20 \div 4 = 5$."

Using tables to perform division

When you and your child worked on learning multiplication, you used tables to illustrate multiplying two numbers. The number of rows and columns corresponds to the two digits in the expression, and the total number of boxes gives you the result. Consider the following expression:

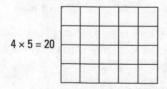

$4 \times 5 = 20$

Tables also work when you must divide one number by another. Here's an example to discuss with your child.

1. **Consider the following expression:**

   $15 \div 3 =$

2. **Explain that because you are dividing 15 by 3, you need to draw a box that has and 3 rows, like the one shown here:**

   Count the total number of boxes. In this case, you have 3.

3. **Draw a vertical line through the boxes as shown here:**

4. **Count the total number of boxes.**

   If the total is not 15, add another line, like this:

5. **Count the number of boxes, and if it is still not 15, add another line:**

6. **Repeat this process until the number of boxes is 15, as shown here:**

7. Count the number of columns, which in this case is 5.

8. Say, "15 ÷ 3 = 5."

Dividing by 1

Earlier, you taught your child that any number multiplied by 1 is that number. Well, it turns out that any number divided by 1 is also that number. Here's an example of a way to talk through this concept.

1. Present the following expressions to your child:

   $5 \div 1 =$ $8 \div 1 =$ $10 \div 1 =$

2. Say to your child, "When we divide a number by 1, we are placing the number of items specified into 1 pile. After we do that and count the items in the pile, it will equal the starting number. In other words, any number divided by 1 is that number."

   If you create 1 pile with 5 straws, you have 5 straws. Likewise, if you create 1 pile of 8 straws, you have 8 straws. Finally, if you create 1 pile of 10 straws, you have 10 straws:

   $5 \div 1 = 5$ $8 \div 1 = 8$ $10 \div 1 = 10$

Knowing that you cannot divide a number by 0

Sometimes you just need to learn the rules, and math is no exception. One of those rules is that you can't divide a number by zero. You just can't do it!

Consider the following expression:

$5 \div 0 =$

If you think in terms of piles of items, this expression tells us to put 0 items in each group, which is impossible if we want to use up our 5 straws. As such, you cannot divide any number by 0. In other words, you can't make a zero pile.

Practicing with division flash cards

Thanks to addition, subtraction, and multiplication flash cards, your child is on their way to becoming a whiz kid. In this section, you build on their skills using division flash cards.

1. Using 3x5 index cards, create the following flash cards:

1÷1	2÷1	3÷1	4÷1	5÷1	6÷1	7÷1	8÷1	9÷1	10÷1
2÷2	4÷2	6÷2	8÷2	10÷2	12÷2	14÷2	16÷2	18÷2	20÷2
3÷3	6÷2	9÷3	12÷3	15÷2	18÷3	21÷3	24÷3	27÷3	30÷3
4÷4	8÷4	12÷4	16÷4	20÷4	24÷6	28÷4	32÷4	36÷4	40÷4
5÷5	10÷5	15÷5	20÷5	25÷5	30÷5	35÷5	40÷5	45÷5	50÷5
6÷6	12÷6	18÷6	24÷6	30÷6	36÷6	42÷6	48÷6	54÷6	60÷6
7÷7	14÷7	21÷7	28÷7	35÷7	42÷6	49÷7	56÷7	63÷7	70÷10
8÷8	16÷8	24÷8	32÷8	40÷8	48÷8	56÷8	64÷8	72÷8	80÷10
9÷9	18÷9	27÷9	36÷9	45÷9	54÷9	63÷9	72÷9	81÷9	90÷10
10÷10	20÷10	30÷10	40÷10	50÷10	60÷10	70÷10	80÷10	90÷10	100÷10

2. Tell your child that they have learned that any number divided by itself (the same number) is 1 and that any number divided by 1 is that number.

As such, they already know the first column of cards — they all equal 1 — and the top row of cards.

3. Have your child start practicing with the first three rows of cards:

1÷1	2÷1	3÷1	4÷1	5÷1	6÷1	7÷1	8÷1	9÷1	10÷1
2÷2	4÷2	6÷2	8÷2	10÷2	12÷2	14÷2	16÷2	18÷2	20÷2
3÷3	6÷2	9÷3	12÷3	15÷2	18÷3	21÷3	24÷3	27÷3	30÷3

4. After your child has mastered these cards, add row 4. Continue to add one row at a time until your child has mastered the cards.

TIP

There's a reason they say, "Practice makes perfect!" You will want to review the division flash cards with your child on a daily basis until they master them. As with other flash cards, you can write the answer on the back of the card so that your child can do a bit of practice on their own.

Completing division problems on a worksheet

Worksheet 16-4 at www.dummies.com/go/teachingyourkidsnewmathfd contains division problems for your child to solve.

REMEMBER

Division is the opposite of multiplication. If your child has mastered the multiplication flash cards, they should find success with this worksheet.

TIP

Another division worksheet, Worksheet 16-5, is on this book's companion website. Download and print several copies of the worksheet with which your child can practice. Time your child as they work. Ideally, they will complete the worksheet in five minutes or less.

Checking Their Work

TIP

Knowing how to check their own work is a valuable skill your child should practice regularly. Your child has learned to check the result of an addition problem using subtraction, and vice versa. In this section, they will learn that they can check the result of their multiplication problems using division, and vice versa. Here's one way to present this idea:

1. **Explain to your child that after performing a multiplication operation, they can check the result using division. Present the following expression to your child:**

   $$6 \times 8 =$$

   Because your child has practiced with their flash cards, they should say 48:

   $$6 \times 8 = 48$$

2. **Tell your child that the way to check the answer is to perform the following division operation:**

   $$48 \div 6 = 8$$

   Say, "In this case, because the numbers match, you know your result is correct."

3. **Present the following expression to your child:**

$54 \div 9 =$

Again, your child should say 6:

$54 \div 9 - 6$

4. **Tell your child, "To check the result of a division operation, you can use multiplication."**

$9 \times 6 = 54$

Again, because the numbers match, your child knows that their results are correct.

IN THIS CHAPTER. . .

» **Mastering multiplication and division**

» **Multiplying two-digit numbers the old school way**

» **Multiplying two-digit numbers using the box method (new math)**

» **Dividing numbers with a remainder**

» **Dividing big numbers**

Chapter **17**

Multiplying and Dividing Large Numbers

I f multiplication and division are opposites, you'd think one of them would be easier. Not so — both take practice. The good news is that by now, your child should have mastered multiplying the numbers through 10 and dividing numbers through 100. In case your child needs to brush up a little, though, this chapter begins with some flash card review. Then I present the old- and new-school multiplication techniques. To wrap up, this chapter brings on division.

TIP

Before you get started, you should have some straws on hand.

Reviewing Multiplication through 10 and Preparing for Timed Tests

TIP

The key to your child's success in multiplying large numbers is their ability to multiply numbers in the range of 1 through 10. Using the multiplication flash cards that you create for Chapter 16, review the flash cards with your child on a daily basis until they have mastered them. Do not move on to multiplying two-digit numbers until your child has mastered the flash cards.

As your child learns multiplication in school, they will be asked to complete timed multiplication worksheets for the numbers 1 through 10. Worksheet 17-1 at this book's companion website (www.dummies.com/go/teachingyourkidsnewmathfd) is an example of what your child might run into as a timed test.

Download and print several copies of the PDF. Time your child as they complete the worksheet. Ultimately, your child should accurately complete the worksheet in 5 minutes or less.

Multiplying Two-Digit Numbers by Single-Digit Numbers (Old School)

To start teaching your child the process of multiplying two-digit numbers, begin with these steps:

1. **Explain to your child, "When you multiply multi-digit numbers, you first multiply the bottom number, in this case 2, by each of the top numbers."**

   Here's an expression to start with:

   ```
    32
   × 2
   ```

2. **Multiply 2 × 2 and ask your child to write the result:**

   ```
    32
   × 2
      4
   ```

3. **Multiply 2 × 3 and ask your child to write the result:**

   ```
    32
   × 2
     64
   ```

   The answer for 32 × 2 is 64.

4. Work through another example with your child:

$$\begin{array}{r} 21 \\ \times\, 4 \\ \hline \end{array}$$

5. Ask your child to repeat the process from Steps 2 and 3 — first multiplying 4 × 1 and then multiplying 4 × 2:

$$\begin{array}{r} 21 \\ \times\, 4 \\ \hline 84 \end{array}$$

6. Explain to your child that depending on the numbers they are multiplying — such as 5 × 2, which is 10 — there are many times when they will need to carry a number and add it to their next result.

Consider the following example:

$$\begin{array}{r} 14 \\ \times\, 5 \\ \hline \end{array}$$

7. To start, have your child multiply 4 × 5, which is 20.

8. Tell your child to write the 0 and then to write a small 2 above the next number to multiply:

$$\begin{array}{r} {}^{2} \\ 14 \\ \times\, 5 \\ \hline 0 \end{array}$$

9. Have your child multiply 5 × 1, which is 5, and then add the 2 that they carried, producing 7:

$$\begin{array}{r} {}^{2} \\ 14 \\ \times\, 5 \\ \hline 70 \end{array}$$

10. Have your child repeat this process for the following expression:

$$\begin{array}{r} 25 \\ \times\, 3 \\ \hline \end{array}$$

They should get:

$$\begin{array}{r} {\scriptstyle 1} \\ 25 \\ \times\,3 \\ \hline 75 \end{array}$$

Worksheet 17-2, found at www.dummies.com/go/teachingyourkidsnewmathfd, is for practicing two-digit multiplication. I've included a few problems here for you to work on with your child. Then ask them to complete the worksheet.

$$\begin{array}{r} 25 \\ \times\,3 \\ \hline \end{array} \qquad \begin{array}{r} 33 \\ \times\,2 \\ \hline \end{array} \qquad \begin{array}{r} 12 \\ \times\,4 \\ \hline \end{array}$$

TIP

Multiplying two-digit numbers requires time and practice. Print several copies of the worksheet and practice them with your child each day until your child masters the process. You should plan for at least of week of practice.

Multiplying two, two-digit numbers

Now that your child understands the process of multiplying a two-digit number by a single-digit number, they are ready to multiply a two-digit number by a second two-digit number. The good news is that the process will start the same: Your child will first multiply the top digits by the ones digit in the second number.

1. **Explain to your child, "You have learned to multiply a two-digit number times a one-digit number. Now, we'll work on multiplying expressions that have two numbers with two digits, such as this."**

$$\begin{array}{r} 12 \\ \times\,31 \\ \hline \end{array}$$

2. **Tell your child that to start, they will multiply the top number times 1:**

$$\begin{array}{r} 12 \\ \times\,31 \\ \hline 12 \end{array}$$

3. **Say to your child, "Now you must multiply the tens digit in the bottom number, which is 3, times the top number. Because you are multiplying the tens digit, you first write a 0 in the spot for the ones digit — that's a rule."**

$$\begin{array}{r} 12 \\ \times\,31 \\ \hline 12 \\ 0 \end{array}$$

4. Say, "Now you can multiply 3 × 12."

$$\begin{array}{r} 12 \\ \times\,31 \\ \hline 12 \\ 360 \end{array}$$

5. Say, "Now you add your two results."

$$\begin{array}{r} 12 \\ \times\,31 \\ \hline 12 \\ 360 \\ \hline 372 \end{array}$$

6. Work with your child to repeat this process for the following expression:

$$\begin{array}{r} 45 \\ \times\,17 \\ \hline \end{array}$$

You should get the following result:

$$\begin{array}{r} {}^{3} \\ 45 \\ \times\,17 \\ \hline 315 \\ 450 \\ \hline 765 \end{array}$$

Download and print Worksheet 17-3 from `www.dummies.com/go/teachingyour kidsnewmathfd` for your child to practice multiplying two-digit numbers. You can assist your child with the few problems I've provided here, and then see if they can complete the worksheet independently.

$$\begin{array}{ccc} 25 & 33 & 12 \\ \times\,23 & \times\,12 & \times\,24 \end{array}$$

TIP

The ability to multiply two-digit numbers is key to being able to multiply larger numbers. Do not move on to larger numbers until your child has mastered the process. Download and print Worksheet 17-4, found at `www.dummies.com/go/teachingyourkidsnewmathfd`, to give your child a tool for practice.

Multiplying numbers through 100

Once your child has the hang of multiplying smaller multi-digit numbers, that skill can be extended to larger numbers with more digits. You and your child can practice advancing this skill with the following steps.

1. Look at the following example with your child:

$$\begin{array}{r} 321 \\ \times\ 412 \\ \hline \end{array}$$

2. Explain to your child, "To start, you will multiply the ones digit in the bottom number by all three digits in the top number."

$$\begin{array}{r} 321 \\ \times\ 412 \\ \hline 642 \end{array}$$

3. Say, "Then, you write your 0 for the tens digit and perform the multiplication."

$$\begin{array}{r} 321 \\ \times\ 412 \\ \hline 642 \\ 3210 \end{array}$$

4. You can state the next step this way: "Because you are going to multiply the hundreds digit, you first write two zeros." Then add the zeroes as shown here:

$$\begin{array}{r} 321 \\ \times\ 412 \\ \hline 642 \\ 3210 \\ 00 \end{array}$$

5. Now have your child perform the multiplication:

$$\begin{array}{r} 321 \\ \times\ 412 \\ \hline 642 \\ 3210 \\ 128400 \end{array}$$

6. **Your child can now add the result of the three numbers:**

```
    321
  × 412
    642
   3210
 128400
 132252
```

7. **Repeat this process for the following expression:**

```
    520
  × 312
```

Your child should get the following solution:

```
    520
  × 312
   1040
   5200
 156000
 162240
```

Worksheet 17-5, located at www.dummies.com/go/teachingyourkidsnewmathfd, includes some expressions for practicing multiplication of three-digit numbers. Help your child with the first couple of problems that I've included here, and then ask them to complete the rest of the worksheet.

```
    251      332
  × 232    × 122
```

Do not move on to division of large numbers until your child has mastered this process. Download Worksheet 17-6 from this book's companion website to give your child more practice.

TIP

Multiplying Numbers Using the Box Method (New Math)

Okay, you've taken a nice tour of Old Math World. It's time to be open to the adventure of some new math! In the following steps, I walk you through performing multiplication on the same types of problems using new math.

1. Present the following problem to your child:

42
× 24

Explain to your child, "You have learned one way to multiply two-digit numbers. Now we will look at a different approach."

2. Draw the following box:

3. Have your child decompose each number they're multiplying, writing the numbers as shown next to the box:

4. Ask your child to multiply 20 × 40, which is 800, and write the result in the upper-left box, as shown here:

5. Repeat this process for 20 × 2, writing the answer in the upper-right box:

6. Repeat this process for 4 × 40 and 4 × 2, writing their results in the lower-left box and lower-right box, respectively:

	40	2
20	800	40
4	160	8

7. The final step is to have your child add the numbers in the boxes by aligning the numbers so that the hundreds, tens, and ones digits align, as shown here:

```
  800
+160
+ 40
+  8
 1008
```

8. Repeat this process for the following expression:

```
  36
× 54
```

Your child should get the following result:

```
 1500
+ 300
+ 120
+  24
 1944
```

Download and print Worksheet 17-7 from www.dummies.com/go/teachingyour kidsnewmathfd to give your child an opportunity to practice multiplication using the box method. Help your child complete the first few problems, and then ask them to complete the worksheet.

Your child can use the box method to multiply larger numbers as well, although using this technique with larger numbers also requires your child to multiply some large numbers as part of the process, such as 100 × 200. I'm including the following example primarily to let you know you can use the box method for large numbers. I would not expect even super whiz kids to multiply such big numbers in their heads at this point.

1. **Present the following expression to your child:**

   103
   × 201

   Then draw the following box and have your child decompose the number into hundreds, tens, and ones as shown here:

	200	0	1
100			
0			
3			

2. **Have your child perform the corresponding multiplication operations, writing their results in the appropriate spots in the box, like so:**

	200	0	1
100	20000	0	100
0	0	0	0
3	600	0	3

3. **Help your child add the numbers within the boxes:**

   20000
   + 600
   + 100
   +   3
   20703

4. **Help your child solve the following expressions using the box method:**

   213   333
   × 402   × 222

You should get the following results:

$$
\begin{array}{r}
80000 \\
+\ 4000 \\
+\ 1200 \\
+\ \ \ 400 \\
+\ \ \ \ 20 \\
+\ \ \ \ \ 6 \\
\hline
85626
\end{array}
\qquad
\begin{array}{r}
60000 \\
+\ 6000 \\
+\ 6000 \\
+\ \ \ 600 \\
+\ \ \ 600 \\
+\ \ \ 600 \\
+\ \ \ \ 60 \\
+\ \ \ \ 60 \\
+\ \ \ \ \ 6 \\
\hline
73926
\end{array}
$$

Revisiting Division through 100 and Preparing for Timed Tests

TIP

The key to your child's success in dividing large numbers is their ability to divide numbers in the range of 1 through 100. Using the division flash cards that you create in Chapter 16, review the flash cards with your child on a daily basis until they have mastered them. Do not move on to dividing two-digit numbers until your child has mastered the single-digit flash cards.

As your child learns division in school, they will be asked to complete timed division worksheets for the numbers 1 through 100. This book's companion website at www.dummies.com/go/teachingyourkidsnewmathfd includes Worksheet 17-8, which is good practice for taking timed division tests.

Download and print several copies of the PDF. Time your child as they complete the worksheet. Ultimately, your child should accurately complete the worksheet in 5 minutes or less.

Recognizing the Second Format for Division

Up to now, I've been using the ÷ symbol for all division problems, but your child needs to be able to recognize the second way division expressions can be written. Show the following expressions to your child and explain that although they look different, they mean the same thing:

$$42 \div 7 \qquad 7\overline{)42}$$
$$35 \div 7 \qquad 7\overline{)35}$$

Worksheet 17-9 on the book's companion web page offers more division practice, but all of the problems are written in this new form. Help your child complete the first few problems, and then ask them to complete the worksheet.

Dividing with a Remainder

When you divide numbers, you don't always get a perfect, whole-number result. Enter the remainder — that little bit that's left over when things don't divide evenly. In this section, your child will learn how to perform division operations that result in a remainder. Visual aids are helpful for learning this concept, so it's time to break out the straws again as you work through these steps:

1. **Hand the straws to be divided to your child and then have them evenly distribute the straws one at a time into the number of piles specified by the second number in the following expression.**

    When they reach a point when they don't have enough straws for each pile, the straws in their hand will be the remainder. Present the following expression to your child:

    $7 \div 3 =$

    Ask your child, "How many straws are you starting with? (7) How many piles should you create? (3)"

2. **Have your child try to evenly divide the straws into 3 piles.**

    They should end up with 1 straw left over.

3. **Explain to your child, "Sometimes we must divide two numbers, and the result is not exact. We call the extra the *remainder*. In this case, you would say that 7 ÷ 3 is 2 remainder 1."**

4. **Consider the following example:**

$$7\overline{)43}$$

Tell your child, "You know that 42 divided by 7 equals 6. You can't multiply 7 times a number to get 43."

Explain that if you subtract 43 – 42, you get 1, so your answer is 6 remainder 1, and they write the result like this:

$$7\overline{)43} = 6 \text{ R}1$$

5. **Repeat this process for the following expressions:**

$$8 \div 5 = \qquad 11 \div 4 = \qquad 7 \div 2 =$$

Your child should get the following results:

$$8 \div 5 = 1 \text{ R}3 \qquad 11 \div 4 = 2 \text{ R}2 \qquad 7 \div 2 = 3 \text{ R}1$$

Help your child work through a few problems on Worksheet 17-10 from www.dummies.com/go/teachingyourkidsnewmathfd before asking them to complete the rest on their own.

Dividing Larger Numbers

Your child will encounter large numbers that they will need to divide. The good news is that the process for dividing large numbers is the same as what they have learned with smaller numbers.

1. **Tell your child, "You have learned how to divide numbers through 100. Now you will learn how to divide larger numbers."**

Consider the following expression:

$$71\overline{)1704}$$

2. **Point to the 1 in 1704 and say, "To start, think about whether 71 can be divided into 1. No, so you then try dividing 17 by 71. Again, that doesn't work because 71 is larger than 17. So, then you try 170. Yes!" Then ask your child, "What's the largest number you can multiply times 71 that is less than or equal to 170?"**

Have your child perform some multiplication operations to determine the number, like these examples:

$$
\begin{array}{ccc}
71 & 71 & 71 \\
\times\,1 & \times\,2 & \times\,3 \\
\hline
71 & 142 & 213
\end{array}
$$

In this case, 3×71 is too large, so you will use 2×71, which is 142.

3. **Perform the division, writing the number 2 above the bar and the result of 2 × 71 below, as shown here:**

$$
\begin{array}{r}
2 \\
71\overline{)1704} \\
142
\end{array}
$$

Say, "Next, you must subtract 170 – 142 and write your result."

$$
\begin{array}{r}
2 \\
71\overline{)1704} \\
142 \\
\hline
28
\end{array}
$$

4. **Explain, "Because we can't divide 71 into 28, we bring down the number 4."**

$$
\begin{array}{r}
2 \\
71\overline{)1704} \\
142 \\
\hline
284
\end{array}
$$

Say, "Now we ask if you can divide 71 into 284. Yes, you can. So, you perform the division by asking, what times 71 goes into 284?" Have your child perform multiplication operations to determine the number, as shown here:

$$
\begin{array}{cccc}
71 & 71 & 71 & 71 \\
\times\,1 & \times\,2 & \times\,3 & \times\,4 \\
\hline
71 & 142 & 213 & 284
\end{array}
$$

In this case, 4×71 is 284. So, you write the 4 above the bar and 4×71 below, as shown here:

```
      24
71)1704
   142
   284
   284
     0
```

The final answer is 24.

Work through another practice problem:

```
35)1365
```

1. **Ask, "Can you divide 35 into 1, 13, or 136?"**

   Your child should know that they'll be dividing into 136 because 1 and 13 are smaller than 35. You can say, "Because you can divide 35 into 136, you ask, 'What is the biggest number I can multiply times 35 that is less than or equal to 136?'"

2. **Have your child try some multiplication next to the problem:**

   ```
    35     35     35     35
   ×1     ×2     ×3     ×4
    35     70    105    140
   ```

   Because $4 \times 35 = 140$ is too large, you will use 3.

3. **Write the 3 above the bar and 35 × 3 below, and do the subtraction, as shown here:**

   ```
        3
   35)1365
      105
       31
   ```

4. **Because you can't divide 35 into 31, you must bring down the next number, 5, like so:**

   ```
        3
   35)1365
      105
      315
   ```

   Explain to your child, "You can divide 35 into 315. What's the biggest number you can multiply times 35 that is less than or equal to 315?"

5. Have your child perform multiplication operations to find the solution:

$$
\begin{array}{ccccc}
35 & 35 & 35 & 35 & 35 \\
\times\,5 & \times\,6 & \times\,7 & \times\,8 & \times\,9 \\
\hline
175 & 210 & 245 & 280 & 315
\end{array}
$$

6. Write 9 above the line and the result of 35 × 9 below, and do the subtraction, like this:

$$
\begin{array}{r}
39 \\
35\overline{)1365} \\
\underline{105} \\
315 \\
\underline{315} \\
0
\end{array}
$$

Because there are no numbers left to bring down, you are done!

Dividing Large Numbers with a Remainder

Remainders don't occur only when you're dividing smaller numbers. Sometimes division problems with large numbers end up with remainders, too. Use the following steps to work through examples of dividing larger numbers that have remainders in the solutions:

1. Consider the following example:

$$23\overline{)717}$$

Ask your child, "Can you divide 23 into 7? How about 71? What's the biggest number you can multiply times 23 that is less than or equal to 71?" Again, have your child try some multiplication to find the number:

$$
\begin{array}{cccc}
23 & 23 & 23 & 23 \\
\times\,1 & \times\,2 & \times\,3 & \times\,4 \\
\hline
23 & 46 & 69 & 92
\end{array}
$$

In this case, $3 \times 23 = 69$, which is less than 71. Write the 3 above the bar and 69 below, and do the subtraction. And because you can't divide 23 into the result, you can also bring down the 7, as shown here:

$$
\begin{array}{r}
3 \\
23\overline{)717} \\
\underline{69} \\
27
\end{array}
$$

2. **In this case, 23 goes into 27 one time. Write that number and subtract the result:**

$$
\begin{array}{r}
31 \\
23\overline{)717} \\
\underline{69} \\
27 \\
\underline{23} \\
4
\end{array}
$$

3. **Because there are no more numbers to bring down, 4 is the remainder, and you write it like this:**

$$
\begin{array}{r}
31\ \text{R}4 \\
23\overline{)717} \\
\underline{69} \\
27 \\
\underline{23} \\
4
\end{array}
$$

4. **Repeat this process for the following expression:**

$$13\overline{)172}$$

Your result should look like this:

$$
\begin{array}{r}
13\ \text{R}3 \\
13\overline{)172} \\
\underline{13} \\
42 \\
\underline{39} \\
3
\end{array}
$$

Download from the companion website and print Worksheet 17-9 so your child can practice division operations that will result in a remainder. Help your child complete the first few problems, and then ask them to complete the worksheet.

Reviewing How to Check Work

When you and your child practiced multiplication and division in Chapter 16, you talked about checking work by doing the opposite operation. In other words, you can check division by multiplying and check multiplication by dividing. Here are some examples you can use to review how to check work for multiplication and division:

$$42 \div 6 = 7 \qquad 7 \times 6 = 42$$
$$5 \times 8 = 40 \qquad 40 \div 8 = 5$$

Explain to your child that when you multiply or divide two-digit numbers, you can still use this method to check your work.

You can explain to your child that to check a division problem that has a remainder, you perform multiplication and then add the remainder to the result. Here's an example of a division problem with a remainder that I've double-checked:

$$
\begin{array}{r}
6 \text{ R}3 \\
7{\overline{\smash{)}45}}
\end{array}
$$

$$
\begin{array}{r}
6 \\
\times\, 7 \\
\hline
42 \\
+\, 3 \\
\hline
45
\end{array}
$$

Chapter **18**

Going Deeper with Charts, Fractions, and Word Problems

f you've ever experienced an income tax audit, you may have tried to explain your math by falling back on rounding errors. This chapter begins by introducing how to round numbers (so maybe your child won't make rounding errors on their income tax returns). Then it revisits charting data. Also, fractions are back again as your child begins adding and subtracting them. The chapter wraps up with a review of word problems. So, roll up your sleeves and let's get started.

Rounding Numbers

This section introduces rounding up and rounding down to the tens and hundreds places, which is a skill that can help them better understand the relationship between numbers. Also, your child can use rounding to perform quick addition.

Rounding numbers through 10

Follow these steps:

1. Say, **"For many math problems, you must determine if a number is closer to 0 or 10 or to 0 or 100. We call this process *rounding a number*."**

   Point to the number 4 on this number line:

   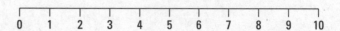

   Ask your child, "Is 4 closer to 0 or to 10?"

2. Tell your child, **"The easy way to tell if a number is closer to 0 or 10 is to split the number line in half, in this case, using the number 5, as you can see here."**

3. Explain to your child, **"In this case, the number 4 is less than 5, which is the halfway point to 10. Because of that, you know that 4 is closer to 0 than it is to 10."**

4. **Point to the number 7 on the number line shown in Step 2.**

   Ask your child, "Is 7 closer to 0 or closer to 10?"

5. **Explain to your child that in this case, because 7 is greater than 5, they would round the number up to 10.**

6. **Repeat this process for the numbers 3 and 8.**

Rounding numbers through 100

When your child is working with larger numbers, they need to know how to round numbers through 100. Use these steps to practice this skill:

1. **Explain to your child, "Just as there will be times when you must round numbers between 0 and 10, you may also round numbers between 0 and 100."**

   Point to the 33 on this number line:

   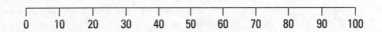

2. **Tell your child that to decide whether to round down to 0 or up to 100, they will use the number 50 as the midpoint between 0 and 100:**

   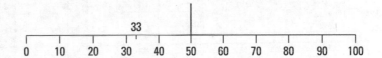

   Explain to your child, "If the number is less than 50, you will round down. Otherwise, if the number is greater than or equal to 50, you will round up."

   Ask your child whether 33 is closer to 0 or to 100.

3. **Repeat this process for the numbers 77 and 29:**

   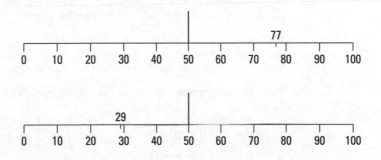

Worksheet 18-1 found at www.dummies.com/go/teachingyourkidsnewmathfd contains a worksheet your child can use to round numbers. Help your child with the first few, and then ask them to complete the worksheet.

Reviewing Charts

Every night on the evening news, we see charts of gas prices going up, stocks going up or down, and team statistics in sports. The ability to read data from charts is a skill your child will use throughout their lifetime. This section gives your child a refresher on gleaning data from charts. Use these steps to help them get started:

1. **Present the following chart to your child:**

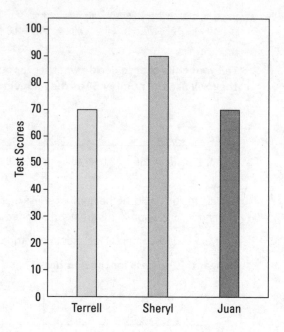

Explain that the chart shows test scores for Terrell, Sheryl, and Juan.

2. **Point to each person's name and score. Ask your child the following questions:**

- Who scored the highest?

- Who scored the lowest?

Charting data

Into each child's life comes a time when a science or social studies project will require charting of some sort of data. That means they need to know how to read the data and turn it into a chart. This exercise helps them practice that skill.

Help your child create a blank chart as shown in the following illustration. Create a vertical bar for the test scores from 0 to 100. Then write the student names beneath the horizontal bar.

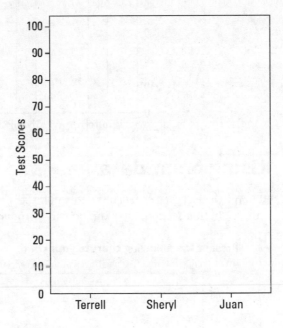

Ask your child to read the following scores:

>> Terrell scored 100

>> Sheryl scored 70

>> Juan scored 80

Help your child to chart the three scores using vertical bars. Your child should produce the following:

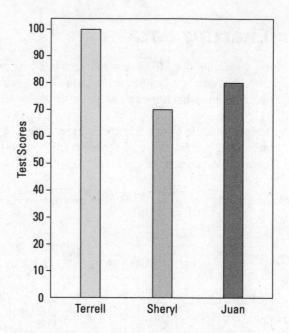

Using point data

When I've discussed charts up to this point, the values have been presented as bar graphs. In this section, I explain how to introduce point data to your child.

1. **Present the following chart to your child:**

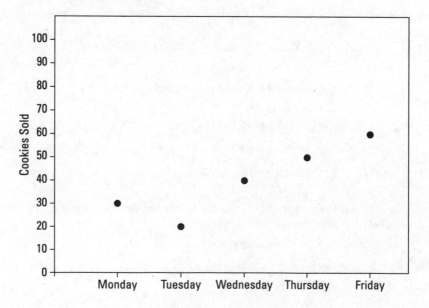

Explain that the chart shows the class cookie sales for Monday through Friday.

2. **Ask your child the following questions:**
   - What day did the class sell the most cookies?
   - What day did the class sell the least cookies?

3. **Draw a trend line on the chart in Step 1 to connect the points, starting with the leftmost point and working to the right.**

   Explain that the line shows them whether sales are going up or down. Ask your child, "Were sales going up at the end of the week?"

4. **Tell your child that they can use this blank chart to show sales for each day of the week:**

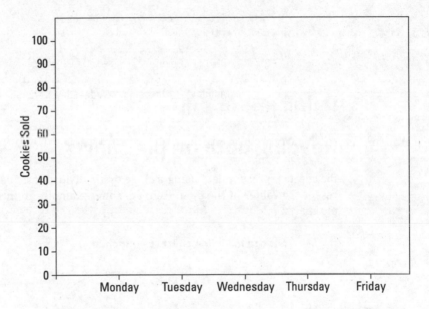

5. **Have your child read the data and then ask them to use points to chart the following data:**
   - On Monday, the class sold 50 cookies.
   - On Tuesday, the class sold 40 cookies.
   - On Wednesday, the class sold 60 cookies.
   - On Thursday, the class sold 65 cookies.
   - On Friday, the class sold 70 cookies.

Your child should produce the following chart:

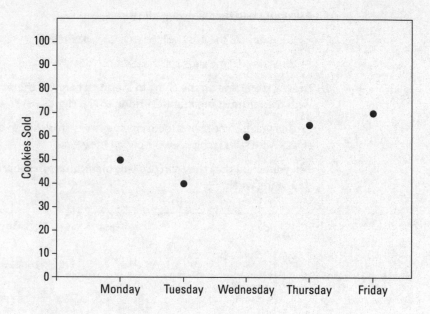

Reading data on pie charts

Pie charts represent data using a circle that's divided into pieces. The pieces represent the values of the data. Here are some examples to introduce your child to pie charts:

1. Present the following chart to your child:

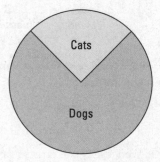

2. Say, "This is a pie chart. You can use it to compare things. In this case, the pie chart compares how many students have cats versus how many have dogs."

In this case, ⅓ of the students have cats, and ⅔ of the students have dogs.

3. **Present the following pie chart to your child:**

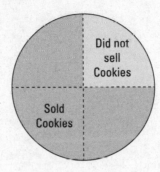

Explain to your child, "This pie chart compares the students in the class who sold cookies to the students in the class who did not."

4. **Ask your child these questions about interpreting the data:**

- What fraction of the students sold cookies?

- What fraction of the students did not sell cookies?

In addition to reading data from a pie chart, your child will need to be able to interpret data by marking it on a pie chart. Use these steps to practice this process:

1. **Tell your child that they are going to create a pie chart that compares how many students read a book versus how many did not.**

2. **On the following blank pie chart, have your child shade and label the chart based on this data:**

- ¾ of the students read a book.

- ¼ of the students did not read a book.

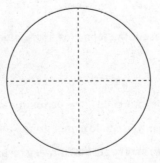

Your child should produce a chart similar to the following:

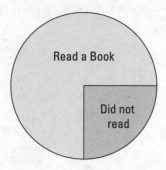

Adding Like Fractions

In this section, your child will learn to add "like" fractions, which means that both fractions have the same denominator (the same number on the bottom). Use these steps to show them the process:

1. **Show the following expression to your child:**

$$\frac{1}{3} + \frac{1}{3} =$$

Say, "When you add fractions, the process is easy if they both have the same denominator — that's the number on the bottom of the fraction. In this case, both fractions have the denominator 3."

2. **Explain to your child, "To add such fractions, you simply add the top numbers and leave the bottom numbers unchanged, as you see here."**

$$\frac{1}{3} + \frac{1}{3} = \frac{2}{3}$$

3. **Present the following expression to your child:**

$$\frac{1}{5} + \frac{2}{5} =$$

4. **Ask your child if the denominators are the same.**

5. **Have your child complete the addition.**

They should get the following result:

$$\frac{1}{5} + \frac{2}{5} = \frac{3}{5}$$

6. **Repeat this process for the following expressions:**

$$\frac{1}{4} + \frac{1}{4} =$$

$$\frac{3}{4} + \frac{2}{4} =$$

$$\frac{1}{2} + \frac{1}{2} =$$

Your child's answers should look like the following:

$$\frac{1}{4} + \frac{1}{4} = \frac{2}{4}$$

$$\frac{3}{4} + \frac{2}{4} = \frac{5}{4}$$

$$\frac{1}{2} + \frac{1}{2} = \frac{2}{2}$$

Worksheet 18-2 at www.dummies.com/go/teachingyourkidsnewmathfd offers practice for adding like fractions. Help your child with the first few, and then ask them to complete the rest of the worksheet.

Subtracting Like Fractions

Subtracting like fractions works similarly to adding them. If the denominators are the same, it's not too tricky. Work through these steps with your child:

1. **Say to your child, "Like addition, subtraction of fractions is easy when the denominators are the same. To subtract one fraction from another, you simply subtract the top numbers, leaving the bottom numbers unchanged."**

Consider the following expression:

$$\frac{2}{3} - \frac{1}{3} =$$

2. **Have your child solve the problem by subtracting the top numbers, as shown here:**

$$\frac{2}{3} - \frac{1}{3} = \frac{1}{3}$$

3. **Repeat this process for the following expressions:**

$$\frac{3}{4} - \frac{1}{4} =$$

$$\frac{3}{5} - \frac{2}{5} =$$

$$\frac{5}{8} - \frac{3}{8} =$$

Your child should get the following answers:

$$\frac{3}{4} - \frac{1}{4} = \frac{2}{4}$$

$$\frac{3}{5} - \frac{2}{5} = \frac{1}{5}$$

$$\frac{5}{8} - \frac{3}{8} = \frac{2}{8}$$

Worksheet 18-3 at www.dummies.com/go/teachingyourkidsnewmathfd includes many practice problems for subtracting like fractions. Help your child with the first few, and then ask them to complete the worksheet.

Reducing Fractions

An improper fraction is a fraction for which the top number is the same or larger than the bottom number. Improper fractions need to be turned into proper fractions, and the process for that is called *reducing*. With the following steps, your child will learn to reduce improper fractions into their proper form.

1. **Have your child solve the following expression:**

$$\frac{1}{2} + \frac{1}{2} =$$

Their answer should be the following:

$$\frac{1}{2} + \frac{1}{2} = \frac{2}{2}$$

2. **Tell your child, "You can think of a fraction as a division operation, which means $\frac{2}{2}$ is the same as saying, '2 divided by 2.'"**

Ask your child to solve 2 divided by 2. Their answer should be 1.

3. **Explain to your child that when they see a fraction with the same top and bottom number, they should replace the fraction with the value 1, as shown in these examples:**

$$\frac{2}{2} = 1 \qquad \frac{3}{3} = 1 \qquad \frac{4}{4} = 1 \qquad \frac{5}{5} = 1$$

4. **Explain how to reduce improper fractions into mixed numbers by saying, "When the top number of a fraction is larger than the bottom number, you need to reduce the fraction into a mixed number. The following fraction needs to be reduced to a mixed number."**

$$\frac{3}{2}$$

Because the top number is larger, your child must reduce the fraction by decomposing the fraction as follows:

$$\frac{3}{2} = \frac{2}{2} + \frac{1}{2}$$

5. **Say, "Because you know that 2/2 is equal to 1, you can rewrite the fraction like this, which is a mixed number."**

$$\frac{3}{2} = \frac{2}{2} + \frac{1}{2} = 1\frac{1}{2}$$

6. **Consider the following expression:**

$$\frac{3}{4} + \frac{2}{4} =$$

Have your child solve the expression. They should get:

$$\frac{3}{4} + \frac{2}{4} = \frac{5}{4}$$

7. **Remind your child that improper fractions like this answer need to be converted to mixed numbers.**

In this case, your child can reduce the fraction as follows:

$$\frac{5}{4} = \frac{4}{4} + \frac{1}{4} = 1\frac{1}{4}$$

8. Ask your child to reduce the following fractions:

$$\frac{4}{3} =$$

$$\frac{7}{5} =$$

$$\frac{10}{7} =$$

Their answers should be the following:

$$\frac{4}{3} = 1\frac{1}{3}$$

$$\frac{7}{5} = 1\frac{2}{5}$$

$$\frac{10}{7} = 1\frac{3}{7}$$

Download and print Worksheet 18-4 from www.dummies.com/go/teachingyour kidsnewmathfd for practice with reducing fractions. Help your child solve the first few, and then ask them to complete the worksheet.

Revisiting Word Problems

It's time to circle back to our old friends: word problems. The examples here incorporate the skills your child has been working on up to this point. Relax, though. There are still no trains leaving the station here!

1. Have your child read the following word problem:

Aiden and Emilia sold 10 cookies. Aiden sold 3. How many cookies did Mary sell?

2. Ask your child the following questions:

- How many cookies did Aiden and Emilia sell?
- How many cookies did Aiden sell?
- Do you need to add or subtract?

3. Write the following expression, and ask your child to solve it:

$10 - 3 =$

4. Have your child read the following word problem:

Dominic has 3 marbles. Chloe has twice as many marbles as Dominic. How many marbles does Chloe have?

5. Ask your child the following questions:

- How many marbles does Dominic have?
- Do you add, subtract, or multiply?

6. Write the following expression, and ask your child to solve it:

$3 \times 2 =$

7. Present the following word problem to your child:

Devin and Mariah are collecting rocks. Devin has 137 rocks and Mariah has 153. How many rocks do Devin and Mariah have in total?

8. Ask your child the following questions:

- How many rocks does Devin have?
- How many rocks does Mariah have?
- Do you add or subtract?

9. Write the following expression, and ask your child to solve it:

$$\begin{array}{r} 137 \\ + 153 \\ \hline \end{array}$$

10. Present the following word problem to your child:

James and Lillian have a pizza. James ate $\frac{2}{5}$ of the pizza, and Lillian ate $\frac{1}{5}$. How much of the pizza did James and Lillian eat?

11. Ask your child the following questions:

- How much pizza did James eat?
- How much pizza did Lillian eat?
- Do you add, subtract, or multiply?

12. Write the following expression, and ask your child to solve it:

$\frac{2}{5} + \frac{1}{5} =$

5

Focusing on Fourth Grade Math

Chapter **19**

Mixing and Matching Operations

This chapter begins with complex expressions. The word *complex* probably makes this sound hard (it isn't), but complex expressions also give you something cool to brag about: "I've been teaching my kid to solve complex expressions." Better yet, when your smart friends on social media post math equations and say, "Can you solve this?" you'll know how to figure out the answer! So, get yourself together (use parentheses if needed), and get ready to have a little math fun!

Working with Complex Expressions

You and your child have worked through adding, subtracting, multiplying, and dividing two numbers at a time, like these:

$$\begin{array}{cccc} 8 & 27 & 3 & 25\overline{)750} \\ \underline{+2} & \underline{-17} & \underline{\times 5} & \end{array}$$

But sometimes, math doesn't stop with just two numbers. Often, your child will need to solve problems that contain multiple values and operators, like these:

$$3+5+7= \qquad 4-2-1= \qquad 3+4-2=$$

Explain to your child that when a problem contains only addition and subtraction, you perform the operations from left to right. Here are some examples to look at with your child:

$$3+5+7= \qquad 4-2-1= \qquad 3+4-2=$$
$$8+7=15 \qquad 2-1=1 \qquad 7-2=5$$

Consider the following expressions:

$$\begin{array}{cccc}
33 & 13 & 14 & 12 \\
+21 & -4 & +17 & +22 \\
\underline{+17} & \underline{-7} & \underline{-15} & \underline{-30}
\end{array}$$

Explain to your child that to solve such problems, they may have to perform two steps, as shown here:

$$\begin{array}{cc cccc ccc}
^{1} & ^{0} & & & & ^{2} & & & \\
33 & 1^{1}3 & 14 & 14 & 3^{1}1 & 12 & 12 & 34 \\
+21 & -4 & +17 & \underline{+17} & \underline{-1\;5} & +22 & \underline{+22} & \underline{-30} \\
\underline{+17} & \underline{-7} & \underline{-15} & 31 & 1\;6 & \underline{-30} & 34 & 4 \\
71 & 2 & & & & & &
\end{array}$$

Also, even though an expression has more than one number, it can still be solved with a number line, as shown in these examples:

$$\begin{array}{c}
33 \\
+\,21 \\
+\,17 \\
\hline
\end{array}$$

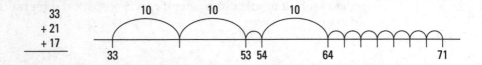

$$\begin{array}{c}
13 \\
-\,4 \\
-\,7 \\
\hline
\end{array}$$

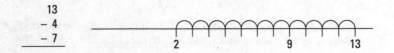

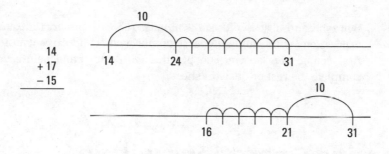

14
+ 17
− 15
_____

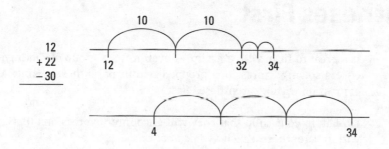

12
+ 22
− 30
_____

Performing Multiplication and Division before Addition and Subtraction

The situation gets slightly more complex when you're mixing multiplication and division with addition and subtraction. The expression has to be solved in a particular order. Look at these expressions with your child:

$2+3\times5=$ $3+6\div2-$

Explain to your child that when an expression has multiple operators, they must perform the multiplication and division operations first:

$2+3\times5=$ $3+6\div2=$
$2+15=17$ $3+3=6$

Ask your child to solve the following expressions:

$13-2\times5=$ $4+2\div2=$ $13+4\div2=$

They should get the following answers:

$13-2\times5=$ $4+2\div2=$ $13+4\div2=$
$13-10=3$ $4+1=5$ $13+2=15$

Worksheet 19-1 at www.dummies.com/go/teachingyourkidsnewmathfd contains expressions with multiple operators that your child can solve. I've included the first couple here so you can practice with your child before you ask them to complete the rest of the worksheet.

$$2+5+1= \qquad 3+4-1=$$

Performing Operations within Parentheses First

It's great to fit in with a group — even if it's a group of math geeks. This lesson covers solving expressions grouped within parentheses. Hint: Always solve the expression within parentheses first.

Explain to your child that they will encounter expressions that group operations within parentheses, such as this one:

$$(3+1)\times 5=$$

Tell your child, "When you see parentheses within an expression, you perform the operations enclosed in the parentheses before you do anything with the other operators." Here's an example:

$$(3+1)\times 5=$$
$$4\times 5=20$$

Help your child complete the following expressions:

$$(3+2)\times 7= \qquad (3-1)\times 4= \qquad (5+2)\times 2=$$

Your results should match what's shown here:

$$(3+2)\times 7= \qquad (3-1)\times 4= \qquad (5+2)\times 2=$$
$$5\times 7=35 \qquad 2\times 4=8 \qquad 7\times 2=14$$

Your child can practice more expressions with parentheses with Worksheet 19-2 at www.dummies.com/go/teachingyourkidsnewmathfd. I've included the first couple of problems here so that you can assist your child if they get stuck. Then they should try to complete the rest of the worksheet on their own.

$$2+(5+1)= \qquad 3+(4-1)=$$

Chapter **20**

Understanding Factors through 100 and Numbers through 1,000,000

Factors. It's a fancy word for a simple concept: two numbers that multiply to produce a specific result. The factors of 2 are 1×2. That was easy!

In this chapter, you and your child revisit some of the multiplication you've already done as you get to know the factors through 100 — in other words, the values you can multiply to get specific values through 100. Along the way, you will learn about prime numbers, such as 3, 5, 7, and 11, for which the only factors are 1 times the number. You might use prime numbers if you want to become an encryption expert (with the hopes of earning millions of dollars). And, speaking of millions, by the end of this chapter, your child will learn to recognize them!

Factoring a Number

You and your child have spent quite a bit of time learning to multiply numbers to produce a result. Now you'll take that concept one step further to understand factoring numbers. Start with these steps:

1. **Consider the following combinations of numbers that multiply to equal 12:**

   $1 \times 12 = 12$ $2 \times 6 = 12$ $3 \times 4 = 12$

   Say to your child, "We call the numbers that you can multiply to produce a specific result *factors*. So the expressions here are factors of 12."

2. **Consider the following factors for the number 36:**

   $1 \times 36 = 36$ $2 \times 18 = 36$ $3 \times 12 = 36$
   $4 \times 9 = 36$ $6 \times 6 = 36$

3. **Ask your child to help you factor the number 24.**

   Say to your child, "To factor a number, start with 1 times the number, which is shown here."

   $1 \times 24 = 24$

4. **Tell your child to try 2 times a number to see if they can get the desired result, like so:**

   $2 \times 12 = 24$

5. **Try to multiply 3 times a number to get the result, like this:**

   $3 \times 8 = 24$

6. **Repeat this process for the number 4:**

   $4 \times 6 = 24$

7. **Then try 5.**

   In this case, 5 does not divide into 24.

8. **Try again with 6.**

   At this point, you can stop because 6 has already appeared in the factor 4×6.

9. **Help your child factor the number 100.**

They should get the following factors:

$$1 \times 100 = 100 \qquad 2 \times 50 = 100 \qquad 4 \times 25 = 100$$
$$5 \times 20 = 100 \qquad 10 \times 10 = 100$$

10. **For more practice, have your child factor the following numbers:**

16	25	40	42	55
80	77	19	33	66

Understanding and Recognizing Prime Numbers

Factors are numbers you can multiply to get a specific result. But for some numbers, the only factor is 1 times the number. We call such numbers *prime numbers*.

Here are some examples of prime numbers:

$$1 \times 2 = 2 \qquad 1 \times 3 = 3 \qquad 1 \times 5 = 5$$
$$1 \times 7 = 7 \qquad 1 \times 11 = 11 \qquad 1 \times 13 = 13$$

Because the only factors for 2, 3, 5, 7, 11, and 13 are 1 times the number, these are prime numbers. There are many prime numbers between 1 and 100.

Ask your child if they can use factoring to identify the prime numbers between 1 and 100. On a blank sheet of paper, write the numbers 1 through 10 and have your child write the corresponding factors. Then, have your child circle the prime numbers (for which the only factor is 1 times the number [2, 3, 5, and 7]).

WHY 1 IS NOT CONSIDERED A PRIME NUMBER

Your child has learned that when the only factor for a number is 1 times the number, that number is a prime number. If you consider the value 1, the only factor is 1×1. However, when the Greeks identified the concept of a prime number, they wanted the number 1 to be treated specially. As such, 1 is not considered a prime number.

Repeat the process of having your child identify factors for the numbers 11 through 20, again circling the prime numbers.

Repeat the process, ten numbers at a time, for the numbers 21 through 100. When your child is done, they should identify the following prime numbers:

2	3	5	7	11
31	37	41	43	47
53	59	61	67	71
73	79	83	89	97

Creating a Factor Tree

Now that your child has learned to identify prime numbers, they are ready to create factor trees, which show the set of prime numbers they can multiply to produce a number. Consider the number 8, which we will draw as the tree's root:

Your child knows that $2 \times 4 = 8$. So you can build the factor tree with 2 and 4, as shown here:

The number 2 is prime, so you can stop that branch. However, the number 4 is not prime, so you continue with that branch:

Again, because each branch is 2, a prime number, you can stop. The factor tree illustrates that $2 \times 2 \times 2 = 8$.

Here's another example with the number 30:

Your child knows that $5 \times 6 = 30$.

The number 5 is prime, so you can stop that branch. The number 6, however, is not prime, so you continue, as shown here:

Because 2 and 3 are prime, you are done. The factor tree illustrates that $5 \times 2 \times 3 = 30$.

Have your child find the prime factors for the number 24. Their result should look like this.

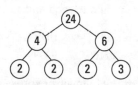

Recognizing Place Values through 1,000,000

Remind your child that they have learned to work with very large numbers and to identify the placeholders for ones, tens, hundreds, and thousands. Use the following steps as a refresher and to introduce the ten-thousands place.

1. **Consider the following number:**

   5267

   Have your child identify and write the ones, tens, hundreds, and thousands digits for 5267 in the appropriate boxes:

2. **Ask your child to say the following number:**

   9999

3. **Ask your child to add the value 1 to that number:**

   $$\begin{array}{r} 9999 \\ +1 \\ \hline 10000 \end{array}$$

4. **Explain to your child that you now have a new placeholder: the ten-thousands place.**

   Have your child write the correct digits in the appropriate placeholder boxes:

   10000

   Tell your child, "In this case we have 0 ones, 0 tens, 0 hundreds, 0 thousands, but 1 ten thousand."

5. **Ask your child to read and say the following numbers:**

10001	10537	10999
13002	21005	99999

6. **Present the following expression to your child:**

   $$\begin{array}{r} 99999 \\ +1 \\ \hline \end{array}$$

   Ask your child to solve the problem. They should get the following result:

   $$\begin{array}{r} 99999 \\ +1 \\ \hline 100000 \end{array}$$

Tell your child that the position of the 1 is the hundred-thousands place.

7. **Have your child place the appropriate digits in each placeholder box, saying out loud the corresponding place (ones, tens, hundreds, and so on):**

100000

8. **Ask your child to read and say the following numbers:**

100001   130000   250000
300010   457153   999999

9. **Present the following expression to your child:**

$$\begin{array}{r} 999999 \\ +1 \\ \hline \end{array}$$

Ask your child to solve the problem. They should get the following result:

$$\begin{array}{r} 999999 \\ +1 \\ \hline 1000000 \end{array}$$

Tell your child that the position of the 1 is the millions place.

10. **Have your child place the appropriate digits in each placeholder box, saying out loud the corresponding place (ones, tens, hundreds, and so on):**

1000000

11. **Ask your child to read and say the following numbers:**

1000001   1300000   1250000
3002010   4571531   9999999

Placing a Comma in Big Numbers

The more digits you add to a number, the harder it is to read. That's why you use commas to help break things up. Use these steps to introduce the concept of adding commas to numbers:

1. Say to your child, "You now know how to recognize numbers bigger than 1 million! To make it easier to read big numbers and to recognize the placeholders, we place commas within large numbers that have four digits or more. Here's an example."

   1,000

2. Tell your child that they figure out where to place a comma within a number by starting at the right side of the number, moving left three digits, and placing the comma, like so:

   $$1\underset{\smile}{0}\underset{\smile}{0}\underset{\smile}{0}$$

   $$1,000$$

3. Explain that if the number is very large, they will place a comma every three digits, as shown here:

   $$1000000$$

   $$1,\underset{\smile}{0}\underset{\smile}{0}\underset{\smile}{0},\underset{\smile}{0}\underset{\smile}{0}\underset{\smile}{0}$$

   $$1,000,000$$

4. Help your child place commas within the following numbers:

1000001	1300000	1250000
3002010	4571531	9999999

   The results should look like this:

1,000,001	1,300,000	1,250,000
3,002,010	4,571,531	9,999,999

Worksheet 20-1 at www.dummies.com/go/teachingyourkidsnewmathfd contains large numbers that your child can use to practice placing a comma every three digits. Help your child solve the first few problems and then ask them to solve the rest.

Chapter **21**

Pressing on with Even and Odd Numbers and Number Patterns

E ven and odd numbers seem easy and one of those things you just know. But that's not the case, because you aren't born just knowing the difference between even and odd numbers; someone along the way taught the concept to you. In this chapter, you do that for your child. Then, you teach your child to identify patterns in number sequences. Relax, you won't have to solve a Fibonacci sequence — for that, there's always Google.

TIP

You may want to have some straws on hand before you start this chapter.

Knowing Even and Odd Numbers

If you look up *odd* in the dictionary, you'll find a definition that's along the lines of "different from what is expected." That would imply that you should expect numbers to be even. That's something you don't usually expect, unless you're playing roulette and have bet on the even numbers.

Tell your child that each number is either an even number or an odd number. Use these steps to explain the difference:

1. **Say, "Even numbers end with 2, 4, 6, 8, or 0."**

   The following numbers are even:

   2   10   14   16   18   20   44   56   88   100

2. **Tell your child that when they count by 2, starting with 2 (2, 4, 6, 8, and so on), they are saying even numbers.**

3. **Explain odd numbers like this: "Odd numbers end with 1, 3, 5, 7, or 9."**

   The following are odd numbers:

   1   3   11   23   35   47   99

4. **Have your child identify whether the following numbers are even or odd:**

   15   37   52   63   72   86   95

Worksheet 21-1 at www.dummies.com/go/teachingyourkidsnewmathfd is a longer list of even and odd numbers that your child can identify. Have your child complete the worksheet by writing and E or O next to a number to identify whether the value is even or odd.

Using your straws, hand your child a group of straws.

Ask your child to divide the straws into two piles, alternating placing straws in each pile.

Your child will end up with either no straws left or one straw left. If your child has no straws left, then you gave them an even number of straws. If they have one left over, then you gave them an odd number. This illustrates that an even number can be evenly divided by 2, whereas an odd number cannot. Ask your child to count the straws and confirm whether the number is odd or even.

Recognizing Number Patterns

As your child's math skills increase, they will start to recognize number patterns. Use these steps to practice identifying number patterns.

1. **Consider the following number pattern that consistently skips numbers:**

   1 2 3 4 _ 6 7 8 9 _11 12 13 14 _16 17 18
   19 _21 22 23 24 _

   Explain to your child that the sequence skips numbers in a pattern. Have them fill in the missing numbers.

2. **Ask your child if they recognize the pattern.**

   If they do not, explain to them that the pattern skips every fifth number.

3. **Ask your child to explain what the pattern would look like if it were to continue for five more numbers.**

4. **Have your child write the next five numbers following the pattern.**

   The result should be the following:

   1 2 3 4 5 6 7 8 9 **10** 11 12 13 14 **15** 16 17 18
   19 **20** 21 22 23 24 **25** 26 27 28 29 **30**

5. **Show the following number sequence to your child:**

   5 10 15 20 _ 30 _ _ 45 50

6. **Ask your child if they see a number pattern, and ask them to fill in the missing numbers.**

   They should get:

   5 10 15 20 **25** 30 **35 40** 45 50

7. **Present the following number sequence to your child:**

   1 2 _4 5 _7 8 _10 11_

8. **Ask your child if they see a pattern. If so, ask them to extend the pattern by six more numbers, which would result in this:**

   1 2 _4 5 _7 8 _10 11_13 14 _16 17_

9. **Ask your child to fill in the missing numbers.**

   They should get:

   1 2 **3** 4 5 **6** 7 8 **9** 10 11 **12** 13 14 **15** 16 17 **18**

10. Show the following number sequence to your child:

_ 2 _ 4 _ 6 _ 8 _ 10 _ 12 _ 14 _ 16 _ 18 _ 20 _

11. Ask your child if they see a pattern, and have them describe it.

12. Have your child extend the pattern for ten more numbers.

They should get:

_ 2 _ 4 _ 6 _ 8 _ 10 _ 12 _ 14 _ 16 _ 18 _ 20 _
22 _ 24 _ 26 _ 28 _ 30

13. Show the following number sequence to your child:

4 8 12 16 20 24 28 32 40 44 48 52

14. Ask your child if they see a pattern, and ask them to extend the patten for three more numbers.

You may need to say, "By how much is each number changing?" They should get:

4 8 12 16 20 24 28 32 40 44 48 52 **56 60 64**

15. Present the following number sequence to your child:

77 70 63 56 49 42 35 28

16. Ask your child if they see a pattern.

You may need to say, "By how much is each number decreasing?"

17. Ask your child to extend the pattern by four numbers.

They should get:

77 70 63 56 49 42 35 28 **21 14 7 0**

18. Present the following number sequence to your child:

1 _ _ 4 _ 6 _ 8 9 10 _ 12 _ 14 15 16 _ 18 _ 20

It's harder, so your child will have to think about a special type of number they learned about in Chapter 20, "Understanding Factors through 100 and Numbers through 1,000,000." (**Hint:** think *prime time.*)

19. Ask your child to fill in the missing numbers, and ask what is unique about the numbers.

You may need to ask your child to factor the numbers that they filled in.

Worksheet 21-2 at www.dummies.com/go/teachingyourkidsnewmathfd is filled with number sequences that your child can complete for practice.

Chapter **22**

Doing Math with Multi-digit Numbers

H opefully, you've had success with the new–math concepts presented in the previous chapters. You might even find yourself thinking that new math isn't so hard.

In this chapter, your child will work with multi-digit numbers — you know, numbers like 10 that have 2 digits. You'll start using old school techniques and then you'll look at the new school approaches. After you get a couple of solutions under your belt, you may start to become a "new math" convert!

This chapter covers a lot of content and is the book's longest chapter. Plan to take appropriate time with each section, not moving on to the next section until your child has mastered the preceding content.

Reviewing Addition with Regrouping

If it's been a while since your child has practiced adding with regrouping, you can use this section as a little reminder of what needs to happen. Otherwise, you can skip on to the next section where the lessons focus on adding more than two numbers.

Use the following examples for some brush-up practice:

1. **Have your child solve this expression:**

   $$\begin{array}{r} 23 \\ +59 \\ \hline \end{array}$$

   Remind your child that $3 + 9 = 12$, so they need to write the 2 and carry the 1. They should get this result:

   $$\begin{array}{r} ^1 \\ 23 \\ +59 \\ \hline 82 \end{array}$$

2. **Have your child complete the following expression:**

   $$\begin{array}{r} 41 \\ +39 \\ \hline \end{array}$$

   They should get this answer:

   $$\begin{array}{r} ^1 \\ 41 \\ +39 \\ \hline 80 \end{array}$$

Adding Three Rows of Multi-digit Numbers

Throughout the day, we often face problems that require us to add several values, such as tracking our spending at the grocery store, keeping score for multiple players in a card game, and so on.

The number of rows you must add doesn't affect the process; the steps will be the same whether you're adding two numbers or three. You must first add the ones column and carry digits to the tens column as necessary. Use these steps to work through some examples:

1. **Present the following problem to your child:**

   ```
     21
   + 31
   + 45
   ```

2. **Have your child start by adding the ones column, as shown here:**

   ```
     21
   + 31
   + 45
      7
   ```

3. **Have your child add the tens column, like so:**

   ```
     21
   + 31
   + 45
     97
   ```

4. **Explain to your child that the numbers they add in the ones column determine whether they need to carry, as well as whether they carry a 10 or 20, or more.**

   Consider this problem:

   ```
     37
   + 28
   + 19
   ```

5. **Ask your child to add the ones column.**

   In this case, they will get 24. That means they should write the 4 in the ones column and carry 20 to the tens column:

   ```
    2
     37
   + 28
   + 19
      4
   ```

6. **Have your child add the tens column:**

```
   2
  37
+ 28
+ 19
  84
```

Worksheet 22-1 at www.dummies.com/go/teachingyourkidsnewmathfd contains problems with three rows of two-digit numbers to be added. I've included one more problem here for you to work through with your child as practice. After doing it, ask them to try completing the worksheet.

```
  12
+ 13
+ 14
```

Adding Numbers through 1,000 (Old School)

Now that your child has mastered addition of two-digit numbers, they can press on to bigger numbers, which proves that the more math you learn, the more math there is to do.

1. **Present the following problem to your child:**

```
  247
+ 162
```

Say to your child, "This problem has hundreds digits. Your process to add the numbers is the same. You will add the ones column, then the tens, and finally the hundreds."

2. **Have your child add the ones column, as shown here:**

```
  247
+ 162
    9
```

3. Have your child add the tens column, writing the 0 and carrying a 1:

$$
\begin{array}{r}
^1 \\
247 \\
+\ 162 \\
\hline
09 \\
\end{array}
$$

4. Have your child add the hundreds column:

$$
\begin{array}{r}
^1 \\
247 \\
+\ 162 \\
\hline
409 \\
\end{array}
$$

Worksheet 22-2 at www.dummies.com/go/teachingyourkidsnewmathfd contains three-digit numbers for addition. I've included one example here for you to work through with your child. Then ask them to complete the worksheet.

$$
\begin{array}{r}
121 \\
+\ 144 \\
\hline
\end{array}
$$

Adding Multi-digit Numbers through 10,000

Okay, the numbers are getting bigger, but the addition process doesn't change. You add a column, carry as necessary, and then add the next column. This process could go on forever!

Transfer the following problem to a piece of paper and work with your child to help them solve it:

$$
\begin{array}{r}
5231 \\
+\ 1358 \\
\hline
\end{array}
$$

Remind your child that even though the following example includes big numbers, the addition process is the same. They start adding at the ones column. Their result should be:

$$
\begin{array}{r}
5231 \\
+\ 1358 \\
\hline
6589 \\
\end{array}
$$

Worksheet 22-3 at www.dummies.com/go/teachingyourkidsnewmathfd contains problems with large numbers to be added. Help your child as necessary to complete the worksheet.

Adding Multi-digit Numbers Using a Number Line (New School)

Even with larger numbers, your child can use number lines for adding. Here's an example of using a number line with two-digit numbers:

1. **Present the following expression to your child:**

   57
   + 33

2. **Draw a number line on a piece of paper and then mark the number 57, as shown here:**

   57

3. **Have your child draw the 3 tens and 3 ones to find the solution to the addition problem:**

   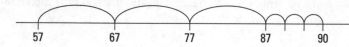

   57   67   77   87   90

4. **Repeat this process for the following expression:**

   47
   + 51

   Draw a number line and mark 47, as shown here:

   47

5. Have your child draw the 5 tens and the 1 one:

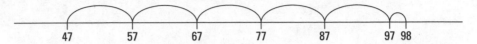

47   57   67   77   87   97 98

Your child can use number lines with three- and four-digit addition problems, too.

1. Present the following problem to your child:

347
+ 231

2. Using a piece of paper, draw a number line and mark 347, as shown here:

347

3. Have your child draw the 2 hundreds:

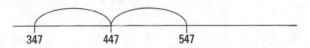

347   447   547

4. Have your child draw the 3 tens and the 1 one:

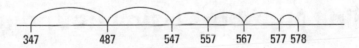

347   487   547   557   567   577 578

5. Repeat this process for the following problem:

553
+ 321

They should get the following result:

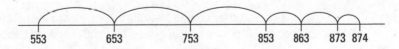

553   653   753   853   863   873 874

Worksheet 22-4 at www.dummies.com/go/teachingyourkidsnewmathfd
contains number lines and expressions for your child to solve. Help your child
with the first few, and then ask them to complete the worksheet.

6. **Present the following four-digit problem to your child:**

3247
+ 2514

Using a piece of paper, draw a number line and mark 3247; then draw the 2 thousands, as shown here:

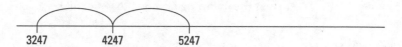

7. **Draw the 5 hundreds:**

8. **Draw the 1 ten and the 4 ones:**

Adding Numbers Using Rounding

Earlier in this chapter, your child solved two-digit addition problems using old school carrying techniques. Another approach to adding such numbers is to use rounding, as I describe here using the following expression:

33
+ 57

1. **With rounding, your child rounds up one of the numbers to its next 10.**

In this case, they'll round 57 to 60, writing next to the problem the value they had to add (in this case 3) to the number 57 to round it to 60:

33
+ 60   *3*

2. Add 33 + 60, as shown here:

$$
\begin{array}{r}
33 \\
+\ 60 \quad \textit{3}\\
\hline
93
\end{array}
$$

3. Subtract the 3 that was added to 57 when it was rounded to 60 to get the final result:

$$
\begin{array}{r}
93 \\
-\ 3 \\
\hline
90
\end{array}
$$

TIP

Did you notice that with this method, your child solved a problem that would require carrying without having to carry?

4. Repeat the process for the following expression:

$$
\begin{array}{r}
34 \\
+\ 27 \\
\hline
\end{array}
$$

Round the number 27 to 30 and write the 3 out to the side:

$$
\begin{array}{r}
34 \\
+\ 30 \quad \textit{3}\\
\hline
\end{array}
$$

5. Solve 34 + 30 and then subtract the 3 to get the final result:

$$
\begin{array}{r}
34 \\
+\ 30 \quad \textit{3}\\
\hline
64 \\
-\ 3 \\
\hline
61
\end{array}
$$

When you deal with three-digit numbers, you must round to the nearest hundred, as shown in this example.

1. Present the following problem to your child:

$$
\begin{array}{r}
349 \\
+\ 274 \\
\hline
\end{array}
$$

Round 274 to 300 by adding 26 and writing the 26 to the side of the expression:

```
    349
+ 300    26
```

2. **Have your child solve 349 + 300:**

```
    349
+ 300    26
    649
```

3. **Have your child subtract the 26:**

```
    349
+ 300    26
    649
-    26
    623
```

Worksheet 22-5 at www.dummies.com/go/teachingyourkidsnewmathfd offers three-digit addition expressions for practice with solving addition operations using rounding. Help your child solve the first few, and then ask them to complete the worksheet.

TIP

Adding numbers using rounding is pretty slick, but it introduces several steps. Remind your child to take their time as they work through the process. When they are done, they should check their work using subtraction as described in Chapter 15.

Subtracting Multi-digit Numbers

You may have guessed that there are several approaches to subtracting multi-digit numbers: carrying, number lines, and rounding. This section covers each technique.

Subtracting multi-digit numbers using borrowing (old school)

Let's start the subtraction process with a little old school regrouping. You know, borrowing from others as necessary — kind of like the national debt. Use these steps to practice this method:

1. **Present the following problem to your child:**

   87
   − 59

   Remind your child that they can't subtract 9 from 7, so they must borrow 10.

2. **Ask your child to solve the problem.**

   They should get the following answer:

   ```
       7
     8¹7
   − 5 9
   ─────
     2 8
   ```

3. **Repeat this process for the following expression:**

   93
   − 77

   Your child should get the following result:

   ```
       8
     9¹3
   − 7 7
   ─────
     1 6
   ```

Worksheet 22-6 at www.dummies.com/go/teachingyourkidsnewmathfd offers some good old school practice of two-digit subtraction. Because your child has previously practiced subtraction with borrowing, see if they can complete the entire worksheet on their own, but provide assistance if needed.

Of course, subtracting larger numbers with three and four digits also sometimes requires borrowing. Here are some examples.

1. **Present this problem to your child:**

   357
   − 268

Remind your child that they can't subtract 8 from 7 and will need to borrow. They should get the following result:

$$\begin{array}{r} {\scriptstyle 2\ \ 14}\\ \not{3}\ \not{8}{}^{1}7\\ -\ 2\ 6\ 8\\ \hline 8\ 9 \end{array}$$

2. **Repeat this process for the following problem:**

$$\begin{array}{r} 414\\ -\ 278\\ \hline \end{array}$$

Your child's result should look like this:

$$\begin{array}{r} {\scriptstyle 3\ \ 10}\\ \not{4}\ \not{1}{}^{1}4\\ -\ 2\ 7\ 8\\ \hline 1\ 3\ 6 \end{array}$$

3. **Present the following problem to your child:**

$$\begin{array}{r} 3217\\ -\ 2183\\ \hline \end{array}$$

4. **Have your child start with the ones column, subtracting 7 – 3:**

$$\begin{array}{r} 3217\\ -\ 2183\\ \hline 4 \end{array}$$

5. **Tell your child that they will need to borrow to subtract 1 – 8 in the tens column, as shown here:**

$$\begin{array}{r} {\scriptstyle 1}\\ 3\ \not{2}{}^{1}1\ 7\\ -\ 2\ 1\ 8\ 3\\ \hline 3\ 4 \end{array}$$

6. **Then, have your child complete the problem.**

They should get this result:

$$\begin{array}{r} {\scriptstyle 1}\\ 3\ \not{2}{}^{1}1\ 7\\ -\ 2\ 1\ 8\ 3\\ \hline 1\ 0\ 3\ 4 \end{array}$$

Subtracting multi-digit numbers using a number line (new school)

Subtracting with a number line works similarly to adding with a number line, but you work in the opposite direction along the line. Use these steps to work through the examples with your child.

1. **Present the following problem to your child:**

   32
   − 21

2. **Draw a number line and mark 32, as shown here:**

   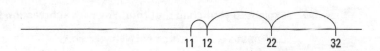

3. **Subtract the 2 tens and then the 1 one, as shown on this number line:**

   ```
   11 12        22        32
   ```

4. **Present the following problem to your child and draw a number line with 347 marked:**

   347
   − 263

5. **Have your child subtract the 2 hundreds, as shown here:**

   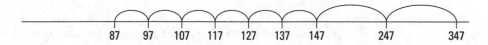

6. **Have your child subtract the 6 tens and 3 ones, like so:**

Download and print Worksheet 22-7 from www.dummies.com/go/teachingyourkidsnewmathfd so that your child can practice solving subtraction problems using a number line. Help your child solve the first few, and then ask them to complete the worksheet.

Solving subtraction problems using rounding

TIP

Earlier you taught your child how to add numbers using rounding. It turns out that you can also use rounding to subtract. Before you get started, understand that subtracting with rounding is a little tricky. Take your time and have your child check their results using addition.

1. **Say to your child, "You have learned to solve addition problems by rounding. Here is an example."**

$$
\begin{array}{ccc}
34 & 34 & 64 \\
+26 & +30 \quad 4 & -4 \\
\hline
 & 64 & 60
\end{array}
$$

2. **Explain to your child that rounding with subtraction is similar.**

   Look at this example problem with your child:

$$
\begin{array}{r}
34 \\
-26 \\
\hline
\end{array}
$$

3. **Have your child round the second number down to the preceding 10, and write the number they subtracted to round down out to the side.**

   In this case, 26 is rounded down to 20, so 6 was subtracted:

$$
\begin{array}{r}
34 \\
-20 \quad 6 \\
\hline
\end{array}
$$

4. **Have your child perform the subtraction for 34 – 20 and then also subtract the 6 that was left over after rounding, like so:**

$$
\begin{array}{r}
34 \\
-20 \quad 6 \\
\hline
14 \\
-6 \\
\hline
8
\end{array}
$$

5. **Repeat this process for the following problem:**

$$
\begin{array}{r}
71 \\
-\ 57 \\
\end{array}
$$

The end result should look like this:

$$
\begin{array}{r}
71 \\
-\ 50 \quad 7 \\
\hline
21 \\
-\ 7 \\
\hline
14 \\
\end{array}
$$

Rounding also works with larger numbers and subtraction. The difference is that you round to the nearest hundred, as in this example.

1. **Present the following problem to your child:**

$$
\begin{array}{r}
712 \\
-\ 575 \\
\end{array}
$$

2. **Explain to your child that because you are using hundreds digits, you round down to 500:**

$$
\begin{array}{r}
712 \\
-\ 500 \quad 75 \\
\hline
212 \\
\end{array}
$$

3. **At this point, you can repeat the process for 212 − 75 to complete the subtraction problem.**

In this case, you round up to 100. Note that because you round *up* rather than down, you *add* 25 to the result of 212 − 100 instead of subtracting as in all the other examples:

$$
\begin{array}{rr}
212 & 212 \\
-\ 75 & -\ 100 \quad 25 \\
& \hline
& 112 \\
& +\ 25 \\
& \hline
& 137 \\
\end{array}
$$

Multiplying Multi-digit Numbers (Old School)

If you just tried rounding to solve addition and subtraction problems, you may feel like you've earned a treat, but there's still more to do. (Of course, you can take a treat break any time you want. This will still be here when you get back.) The good news is that multiplication is back, but you're starting the old school way, which may feel more familiar to you.

Follow these steps to introduce this concept:

1. **Present the following problem to your child:**

    $$
    \begin{array}{r}
    65 \\
    \times\, 34 \\
    \hline
    \end{array}
    $$

2. **Have your child start by multiplying 4 × 5, which is 20. Have them write the 0 and carry the 2:**

    $$
    \begin{array}{r}
    {\scriptstyle 2} \\
    65 \\
    \times\, 34 \\
    \hline
    0
    \end{array}
    $$

3. **Have them multiply 4 × 6, which is 24, add the 2 to produce 26, and write the result as shown here:**

    $$
    \begin{array}{r}
    {\scriptstyle 2} \\
    65 \\
    \times\, 34 \\
    \hline
    260
    \end{array}
    $$

4. **Move on to multiplying the tens digit.**

    First, write a 0 as a placeholder in the ones column, like this:

    $$
    \begin{array}{r}
    {\scriptstyle 2} \\
    65 \\
    \times\, 34 \\
    \hline
    260 \\
    0
    \end{array}
    $$

5. Have your child multiply 3 × 5, which is 15, write the 5 in the hundreds column, and carry the 1:

```
    1
   65
 × 34
  260
   50
```

6. Have your child multiply 3 × 6, which is 18, add the carried 1 to produce 19, and write the result, as shown here:

```
    1
   65
 × 34
  260
 1950
```

7. Complete the problem by having your child add the two numbers to get the result:

```
    1
   65
 × 34
  260
 1950
 2210
```

8. Repeat this process for the following expression:

```
   71
 × 19
```

They should get the following result:

```
   71
 × 19
  639
  710
 1349
```

Have your child practice some more two-digit multiplication with Worksheet 22-8 from www.dummies.com/go/teachingyourkidsnewmathfd. Your child may need help with the first few, but then they'll probably be ready to complete the rest of the worksheet on their own.

Multiplication with three-digit numbers works similarly. By the time you get to multiplying the number in the hundreds place, though, you need two zeroes as placeholders. Here's an example to work through.

1. **Present the following expression to your child:**

   ```
     217
   × 135
   ```

2. **Have your child multiply the ones digit:**

   ```
        3
      217
   × 135
     1085
   ```

3. **Before they multiply the tens digit, have your child write a 0 as a place-holder in the ones column; then they can do the multiplication with the tens digit, as shown here:**

   ```
      217
   × 135
     1085
     6510
   ```

4. **To multiply the hundreds digit, your child needs to write two zeroes and then perform the multiplication, like so:**

   ```
      217
   × 135
     1085
     6510
   21700
   ```

5. **Now your child can add the results of the three numbers to arrive at the final result:**

   ```
      217
   × 135
     1085
     6510
   21700
   29295
   ```

6. **Repeat this process for the following problem:**

> 247
> × 213

Your child should get this result:

> 247
> × 213
> 741
> 2470
> 49400
> 52611

Multiplying Multi-digit Numbers Using the Box Method (New School)

TIP

In this section, you will teach your child to multiply large numbers using the box method. I like the box method. It works. The challenge is that your child must be able to multiply numbers such as 40 × 50, although they don't need to do that math in their head. They can use a scratch piece of paper for those computations so they can write the results in the box.

Use the following problem for this example:

> 47
> × 56

1. **Create the following box, which decomposes the numbers into tens and ones:**

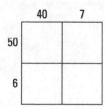

2. Have your child multiply the tens and ones digits (using scrap paper, if necessary) and write the results in the box, as shown here:

	40	7
50	2000	350
6	240	42

3. To finish, your child can add the four numbers to get the final result:

```
  2000
+  350
+  240
+   42
  2632
```

4. Repeat this process for the following problem:

```
  66
× 17
```

The boxes for this problem should look like this:

	60	6
10	600	60
7	420	42

The addition problem to get the final result should look like this:

```
   1
  600
+ 420
+  60
+  42
 1122
```

Worksheet 22-9 at www.dummies.com/go/teachingyourkidsnewmathfd is a multiplication worksheet with problems your child can solve using the box method. Work through one or two with your child, and then ask them to complete the worksheet.

Dividing Multi-digit Numbers

TIP

In this section, your child will learn to divide large numbers using long division. Your goal is to get your child familiar with the process. Long division takes practice, and your child will work on this skill more in sixth grade.

1. **Present the following problem to your child:**

$$25\overline{)325}$$

Ask your child, "Can you divide 25 into 3? How about into 32?" Ask your child to think about what is the largest number they can multiply times 25 that is less than or equal to 32. In this case, the answer is 1, so write the 1 above the bar and the result of 25×1 below:

$$
\begin{array}{r}
1 \\
25\overline{)\,325} \\
-25 \\
\hline
7
\end{array}
$$

2. **Because you can't divide 25 into 7, you bring down the 5.**

In this case, $25 \times 3 = 75$, so write the 3 above the bar and the result of 25×3 below, as shown here:

$$
\begin{array}{r}
13 \\
25\overline{)\,325} \\
-25 \\
\hline
75 \\
75 \\
\hline
0
\end{array}
$$

Because the result is 0, your child has reached the final result.

3. **Repeat this process for the following problem:**

$$17\overline{)374}$$

You should get:

$$\begin{array}{r} 22 \\ 17\overline{)\ 374} \\ -34 \\ \hline 34 \\ -\ 34 \\ \hline 0 \end{array}$$

Download and print Worksheet 22-10 from www.dummies.com/go/ teachingyourkidsnewmathfd so your child can do some additional division practice. Help them as necessary to complete the worksheet.

Chapter 23

Going Deeper with Fractions

They're back! It's time to revisit our good ol' friends, fractions.

In this chapter, you remind your child how to add and subtract fractions. Along the way, your child learns about mixed numbers and improper fractions and how to reduce them. The good news is that after you get past the unfamiliar terms (such as *denominator* and *numerator*), the processes are straightforward.

Mastering fractions takes practice, so this chapter reviews key concepts this book introduced in earlier grade levels. The review should build your child's confidence while providing the foundation for this chapter's new concepts.

If your child asks you why they need to learn this level of detail about fractions, you might explain that it comes in handy when, for example, they're cooking Thanksgiving dinner and extending Grandma's gravy recipe for 4 people that requires ½ cup of flour, and they need to figure out how to make enough gravy for 12 guests!

Reviewing Adding Fractions

Present your child with the following expression:

$$\frac{1}{3} + \frac{1}{3} =$$

Remind your child that when the fractions have the same *denominator* (the bottom number), the addition process is straightforward because they simply add the top numbers and leave the bottom numbers the same:

$$\frac{1}{3} + \frac{1}{3} = \frac{2}{3}$$

Ask your child to add the following fractions:

$$\frac{1}{4} + \frac{2}{4} =$$

They should get the following result:

$$\frac{1}{4} + \frac{2}{4} = \frac{3}{4}$$

Worksheet 23-1 at www.dummies.com/go/teachingyourkidsnewmathfd is filled with expressions, which your child can use to practice adding like fractions. Help your child with the first few, and then ask them to complete the worksheet.

Reviewing Subtracting Fractions

The previous section reminded you that *like fractions* have the same denominator (bottom number), and that adding them is easy. In this lesson, you'll find that the same is true for *subtracting* like fractions.

Present the following expression to your child:

$$\frac{3}{4} - \frac{1}{4} =$$

Subtracting two fractions that have the same bottom number (denominator) works the same way as when you add them: You simply subtract the numerators (top numbers) and leave the denominators (bottom numbers) unchanged, like this:

$$\frac{3}{4} - \frac{1}{4} = \frac{2}{4}$$

Download and print Worksheet 23-2 from www.dummies.com/go/teachingyourkidsnewmathfd to give your child an opportunity to get the hang of subtracting fractions. Help them with the first few, and then turn them loose on the rest of the worksheet.

Revisiting Fractions and Mixed Numbers

Explain to your child that they have learned to perform various operations with fractions. Say, "Most of the fractions you've worked with in this chapter have been portions of one whole, like these."

$$\frac{1}{2} \quad \frac{1}{3} \quad \frac{2}{5} \quad \frac{1}{4}$$

Sometimes, though, fractions will be preceded by a whole number, such as these, which we call *mixed fractions*:

$$2\frac{1}{2} \quad 3\frac{1}{3} \quad 4\frac{2}{5} \quad 1\frac{1}{4}$$

You can explain that you call them mixed fractions because they have a whole number and a fractional part.

When it's time to add or subtract expressions that include mixed numbers, you have to take some extra steps to get to the solution. Here's a way to explain the process to your child:

1. **Tell them that to perform arithmetic operations such as addition and subtraction with expressions that include mixed numbers, you must convert the mixed numbers into fractions.**

   Consider the following mixed number:

   $$1\frac{1}{4}$$

   It looks like this:

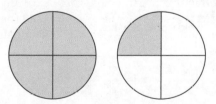

2. Say to your child, "In this case, the fraction is fourths (the denominator is 4), so we can replace the 1 with $\frac{4}{4}$."

$$1\frac{1}{4} = \qquad \frac{4}{4} + \frac{1}{4} =$$

3. Ask your child to add the fractions.

They should get:

$$\frac{4}{4} + \frac{1}{4} = \frac{5}{4}$$

Have your child look at the original illustration of the fraction to confirm the mixed number has $\frac{5}{4}$.

4. Present the following mixed number:

$$2\frac{1}{3}$$

It looks like this:

5. Tell your child, "To determine how many thirds 2 is equal to, you can multiply 2 times the fraction's bottom number (3):"

$$2\frac{1}{3} = \qquad \frac{6}{3} + \frac{1}{3} = \frac{7}{3}$$

Again, have your child review the fraction visualization and confirm the fraction has $\frac{7}{3}$.

6. Repeat this process for the following mixed number:

$$3\frac{1}{4}$$

which looks like this:

The result is:

$$3\frac{1}{4} = \frac{12}{4} + \frac{1}{4} = \frac{13}{4}$$

Have your child review the original fraction visualization and confirm the fraction has $\frac{13}{4}$.

Worksheet 23-3 at www.dummies.com/go/teachingyourkidsnewmathfd offers some good practice for converting mixed numbers to fractions. I've included a couple of problems here so that you can walk your child through them. After doing these additional examples, encourage your child to complete the worksheet on their own.

$$1\frac{3}{4} = \qquad\qquad 2\frac{4}{5} =$$

Reducing Improper Fractions

If you are invited to England to speak with the queen, it's important to be proper. Turns out that it's also important for fractions to be proper. An *improper fraction* is one in which the numerator (top number) is larger than the denominator (bottom number). Use the following steps to help your child understand how to recognize improper fractions and convert them to mixed numbers:

1. **Explain to your child, "When you perform arithmetic operations with fractions, there will be times when you get a result such as this."**

   $$\frac{7}{2}$$

   Explain that because the top number (7) is larger than the bottom number (2), you call this fraction an improper fraction.

2. **To determine the corresponding mixed fraction, you divide the top number by the bottom number, and create the mixed fraction by writing the result as the fraction's whole part, with the remainder becoming the numerator for the fraction.**

3. **Repeat this process for the following improper fraction:**

   $$\frac{7}{2}$$

You should get:

$$\frac{7}{2} = 2\overline{)7} = 2\overline{)\ \ 7}^{\ 3\ R1} = 3\frac{1}{2}$$
$$\underline{-6}$$
$$1$$

Converting improper fractions to mixed fractions takes some getting used to. Use Worksheet 23-4 from www.dummies.com/go/teachingyourkidsnewmathfd to help your child become accustomed to working through the steps for this process. Help your child with the first few, and then ask them to complete the worksheet.

Reducing Fractions

Explain to your child that they have seen many equivalent fractions such as the following:

$$\frac{1}{2} \qquad \frac{2}{4} \qquad \frac{3}{6}$$

Although the fractions are equal, the preferred way is to represent the fraction with the smallest numerator (top number). We call the process of finding the best representation of a fraction *reducing the fraction*.

1. Say to your child, "When we work with fractions, there are times when we have fractions that we can write in a better way. We call the process of putting the fractions in a simpler form *reducing the fraction*."

   Consider the following fraction:

   $$\frac{2}{4}$$

2. Have your child shade in ¾ of the following circle:

3. **Ask your child to tell you how much of the circle is filled in.**

It may help you to prompt them by saying, "Is one-half filled in?"

4. **When your child recognizes that one-half of the circle is filled in, explain that two-fourths is equal to one-half and show them this expression:**

$$\frac{2}{4} = \frac{1}{2}$$

In other words, you can say, "The fraction ¾ reduces to ½."

Tell your child that the good news is that they can learn to reduce fractions without having to draw them. Here's an example to work through without using an illustration.

1. **Present the following fraction to your child:**

$$\frac{3}{6}$$

2. **Tell your child, "To reduce this fraction, start by asking yourself whether you can divide the numerator into the denominator?"**

In this case, 6 can be divided by 3.

3. **Explain that if you can divide the numerator into the denominator, you can reduce the fraction by dividing both numbers by the numerator, as follows:**

$$\frac{3 \div 3}{6 \div 3} = \frac{1}{2}$$

4. **Present the following fraction to your child, and ask if the numerator can divide into the denominator:**

$$\frac{2}{10}$$

5. **Again, you can, so divide both numbers by 2, like so:**

$$\frac{2 \div 2}{10 \div 2} = \frac{1}{5}$$

6. **Present the following fraction to your child:**

$$\frac{4}{6}$$

7. **Ask your child, "Can you divide 6 by 4?"**

In this case, they can't.

8. **Explain to your child that dividing the numerator into the denominator isn't the only way to reduce a fraction.**

Another way is to see if both numbers can be divided by 2. In this case, they can, so the fraction can be reduced like this:

$$\frac{4 \div 2}{6 \div 2} = \frac{2}{3}$$

9. **Present the following fraction to your child:**

$$\frac{3}{9}$$

10. **Have your child first test whether the numbers are divisible by 2.**

In this case, they are not.

11. **Have your child examine whether both numbers are divisible by 3.**

In this case, they are, so the fraction can be reduced like this:

$$\frac{3 \div 3}{9 \div 3} = \frac{1}{3}$$

12. **Present the following fraction to your child:**

$$\frac{4}{9}$$

13. **Ask them if both numbers are divisible by 2.**

They are not.

14. **Ask if both are divisible by 3.**

They are not.

15. **Ask whether both numbers are divisible by 4.**

Again, they are not.

Because the last number you tested is the same as the top number, you can stop. The fraction $\frac{4}{9}$ cannot be reduced.

Like anything else in math, practice makes perfect when reducing fractions. Download and print Worksheet 23-5 from www.dummies.com/go/teaching yourkidsnewmathfd so your child can work on this skill. Help your child with the first few problems, and then ask them to complete the worksheet.

Solving Word Problems That Use Fractions

Word problems get a bad rap — most people don't ride on trains and therefore can't relate to what the word problems are asking. But word problems are truly applicable to real life. That's why my word problems talk about things like pizza. We can all relate to that!

Practice your child's fraction skills with these word problems.

1. **Present the following word problem to your child:**

   Kai has ¾ of a pizza. He eats ¼ of it. How much pizza is left?

   Ask your child:

   - How much pizza did Kai start with?
   - How much pizza did he eat?
   - Do you need to add or subtract?

2. **Write the following expression for your child to solve:**

   $$\frac{3}{4} - \frac{1}{4} =$$

3. **Make sure your child reduces the result.**

   They should get the following result:

   $$\frac{3}{4} - \frac{1}{4} = \frac{2}{4} = \frac{1}{2}$$

4. **Present the following word problem to your child:**

   Mateo has ¼ glass of soda. Keziah has ²⁄₄ glass of soda. How much soda do Mateo and Keziah have?

Ask your child:

- How much soda does Mateo have?
- How much soda does Keziah have?
- Do you need to add or subtract?

5. **Write the following expression for your child to solve:**

$$\frac{1}{4} + \frac{2}{4} =$$

They should get the following result:

$$\frac{1}{4} + \frac{2}{4} = \frac{3}{4}$$

6

Advancing to
Fifth Grade Math

Chapter **24**

Interpreting Mathematical Expressions

f you've ever tried to create a budget, you've worked with expressions that have multiple operators: rent + food + entertainment. You've also probably compared expressions: rent + food + entertainment < income + bank balance! In this chapter, you teach your child how to solve such expressions. The good news is that solving these expressions is straightforward. The bad news is that new math doesn't help budgeting.

Solving Problems Using Multiple Operations

By now, your child has learned to solve many types of problems that use addition, subtraction, multiplication, division, and even fractions. Tell your child that they are now going to learn to solve complex expressions that use several of these operations. Here's an example:

$$5 + 7 + 4 - 3 =$$

In Chapter 19, you and your child worked through the rules for solving expressions with multiple operations — for example, when you have only addition and subtraction as in this example, you perform the operations from left to right. Have your child solve the problem. They should get the following result:

$$5 + 7 + 4 - 3 = 13$$

Here's another example that throws multiplication into the mix:

$$5 + 6 + 6 \times 2 =$$

There's a second rule that applies to expressions like this: multiplication and division operations are done first. So this example can be broken into two steps, like so:

$$5 + 6 + 6 \times 2 =$$
$$5 + 6 + 12 = 23$$

Worksheet 24-1 at www.dummies.com/go/teachingyourkidsnewmathfd includes problems with multiple operators. Help your child with the first few, and then ask them to complete the rest.

Solving Problems with Multi-digit Numbers

This lesson continues with complex expressions that use multiple operators. The difference is that the numbers are bigger.

1. **Present this problem to your child:**

    $$35 + 17 + 14 - 23 =$$

2. **Say to your child, "Because this expression has only addition and subtraction, you will perform the operations from left to right."**

   Suggest to your child that they may want to break down the expression and solve it in two parts to make it easier. They can do the addition operations and then the subtraction operation, like so:

   $$
   \begin{array}{cc}
   35 & 66 \\
   +17 & \underline{-23} \\
   \underline{+14} & 43 \\
   66 &
   \end{array}
   $$

3. **Present the following problem to your child:**

   $$23 - 22 + 16 \times 22 =$$

4. **Remind your child that they should do the multiplication first and then work through the subtraction and addition from left to right, as shown here:**

   $$
   \begin{array}{cc}
   22 & 23 \\
   \underline{\times 16} & \underline{-22} \\
   132 & 1 \\
   \underline{220} & \underline{+352} \\
   352 & 353
   \end{array}
   $$

The expressions in Worksheet 24-2 from www.dummies.com/go/teaching yourkidsnewmathfd combine multi-digit numbers and multiple expressions. Help your child with the first few, reminding them of the proper order for performing the different operations, if necessary. Then ask them to try to complete the worksheet.

Solving Problems Grouped by Parentheses

Let's not forget that there's one more thing that can influence the order for solving an expression. That's right! Parentheses are back!

In Chapter 19, you taught your child that when they see an expression with parentheses, they solve the expression within the parentheses first. Here's an example to work through as a reminder:

$$3 + 2 \times (2 + 3) =$$

The operations grouped by the parentheses need attention first. So, to solve the problem, you work through it like this:

$$3 + 2 \times (2 + 3) =$$
$$3 + 2 \times 5 =$$
$$3 + 10 = 13$$

With Worksheet 24-3 from www.dummies.com/go/teachingyourkidsnewmathfd, your child can practice their skills at solving expressions with multiple operators: parentheses first, multiplication and division second, addition and subtraction (and working from left to right) third. Work through the first few problems with your child, and then see how they fare when they go solo.

Completing Equivalent Expressions

The meaning of *equivalent expressions* is right there in the name: expressions that are equivalent. Another way you can say it is, "expressions that have the same result." (See if you can figure out a way to work the words "equivalent expression" into your next business meeting.) In the following lesson, your child will learn to create equivalent expressions.

1. **Present the following expression to your child:**

   $$3 + 4 = 4 + \underline{}$$

   Your child may be able to quickly solve the problem and tell you the answer is 3.

2. **If your child does not immediately recognize the answer, have them work through the expression like this:**

   $$3 + 4 = 4 + \underline{}$$
   $$7 = 4 + \underline{}$$

3. **Ask your child, "What number must you add to 4 to equal 7?"**

   Seeing the expression broken down into these steps should help them recognize that $3 + 4$ is equivalent to $4 + 3$.

4. **Present the following problem to your child:**

   $$3 + 7 = \underline{} + 5$$

   Your child may be able to solve this problem in their head.

5. **If it seems like your child needs some assistance visualizing how to work through this problem, have them complete the first expression:**

$$3 + 7 = \underline{} + 5$$
$$10 = \underline{} + 5$$

6. **Ask your child, "What number do you add to 5 to equal 10?"**

7. **Present the following expression to your child:**

$$35 + 43 = 37 + \underline{}$$

8. **Ask your child to solve the problem.**

Suggest that they solve the expression on the left first:

$$35 + 43 = 37 + \underline{}$$
$$78 = 37 + \underline{}$$

9. **Ask your child, "What number do you need to add to 37 to equal 70?"**

Your child can answer the question using a number line, or they can subtract the number 37 from 78:

$$\begin{array}{r} 78 \\ -37 \\ \hline 41 \end{array}$$

Your child can use Worksheet 24-4 from www.dummies.com/go/teaching yourkidsnewmathfd to practice solving equivalent expressions. Help your child with the first few, and then ask them to complete the worksheet.

Revisiting Expressions with Parentheses

Your child has learned that when an expression contains different operations, they perform multiplication and division first, from left to right, followed by addition and subtraction. Expressions also often group operations within parentheses, like this:

$$(3+1) \times 5 =$$

Explain that when an expression contains parentheses, your child first solves the operation within the parentheses, followed by multiplication and division, and finally addition and subtraction. In the case of the previous expression, the result would be the following:

$$(3+1) \times 5 =$$
$$4 \times 5 = 20$$

Present the following expression to your child to solve:

$$3 \times (4+1) =$$

They should start with the operation within the parentheses and then perform the multiplication:

$$3 \times (4+1) =$$
$$3 \times 5 = 15$$

Worksheet 24-5 at www.dummies.com/go/teachingyourkidsnewmathfd contains expressions that use parentheses. Help your child with the first few, and then have them complete the rest.

Comparing Expressions Using >, <, and =

If equivalent expressions are the same (or equal), then inequivalent expressions aren't the same — meaning, their results are different. Sometimes, you compare expressions to see whether they're equal or one is greater than the other. You can use the equal sign (=), greater-than sign (>), and less-than sign (<) in these comparisons.

Here are some examples.

1. **Present the following expression to your child, explaining that they will solve the two expressions and then place the greater-than sign (>), less-than sign (<), or equal sign (=) between the brackets:**

    $$3+2 \begin{bmatrix} \end{bmatrix} 4+1$$

    In this case, they should get:

    $$3+2 \begin{bmatrix} \end{bmatrix} 4+1$$
    $$5 \begin{bmatrix} = \end{bmatrix} 5$$

2. **Repeat this process for the following expression:**

$$33 + 44 \; [\;\;] \; 17 \times 5$$

Your child's result should be this:

$$33 + 44 \; [\;\;] \; 17 \times 5$$
$$77 \; [<] \; 85$$

Download and print Worksheet 25-6 from www.dummies.com/go/teaching yourkidsnewmathfd so your child can practice comparing expressions using >, <, and =.

Chapter **25**

Revisiting Fractions, One More Time

ood news! This is the last chapter on fractions! Unfortunately, this chapter introduces more terms, such as *like fractions* and *unlike fractions*. However, the terms make the concepts sound harder than they are! By the end of this chapter, your child will have mastered fractions and the operations you perform with them! You may even start to like fractions — at least the "like fractions."

Comparing Fractions

You and your child have covered a wide range of operations with fractions. In this lesson, you are going to discover an easy way to compare fractions.

1. **Ask your child to examine the following fractions and to use the greater-than, less-than, or equal symbol as appropriate:**

$$\frac{1}{4} \quad \frac{3}{4} \qquad \frac{1}{2} \quad \frac{1}{2} \qquad \frac{3}{5} \quad \frac{4}{5} \qquad \frac{7}{8} \quad \frac{5}{8}$$

2. Say to your child, "When the denominators (the fractions' bottom numbers) are the same, comparing fractions is easy. Often, however, we must compare fractions for which the denominators are different, such as the following."

$$\frac{1}{3} \quad \frac{1}{2}$$

3. Explain to your child that one way to compare fractions is to draw them, as shown here:

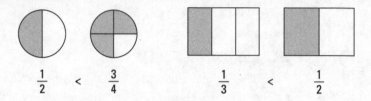

$$\frac{1}{2} \quad < \quad \frac{3}{4} \qquad\qquad \frac{1}{3} \quad < \quad \frac{1}{2}$$

4. Explain that to compare the fractions numerically, you must convert the fractions, so that they have the same denominator.

To do so, you multiply one fraction's denominator by the other fraction's denominator. In this case, the denominator for ⅓ is 3, so you multiply ½ by ³⁄₃. Likewise, the denominator for ½ is 2, so you multiply ⅓ by 2/2:

$$\frac{1}{3} \quad \frac{1}{2} \qquad \frac{1}{3}\times\frac{2}{2}=\frac{2}{6} \qquad \frac{1}{2}\times\frac{3}{3}=\frac{3}{6}$$

5. Say to your child, "After you convert the fractions so that they are like fractions (with the same denominator), you can see that ³⁄₆ (which is ½) is greater than ²⁄₆ (which is ⅓).

Here's another example to work on with your child.

1. Ask your child to consider the following fractions:

$$\frac{3}{5} \quad \frac{4}{6}$$

2. Ask your child to convert the fractions to a common denominator:

$$\frac{3}{5}\times\frac{6}{6}=\frac{18}{30} \qquad\qquad \frac{4}{6}\times\frac{5}{5}=\frac{20}{30}$$

3. By comparing the fractions with the same denominators, you can see that ⁴⁄₆ is greater than ³⁄₅.

Worksheet 25-1 at www.dummies.com/go/teachingyourkidsnewmathfd offers plenty of pairs of unlike fractions for your child to compare using >, <, and =. Help your child complete the first few, and then ask them to complete the worksheet.

Adding Unlike Fractions

Remind your child that adding fractions that have the same denominator is a simple process of adding the numerators and keeping the denominator the same. Here are some examples to practice with:

$$\frac{1}{3} + \frac{1}{3} = \qquad \frac{1}{4} + \frac{2}{4} = \qquad \frac{1}{5} + \frac{3}{5} =$$

Adding unlike fractions is similar to making comparisons between fractions with different denominators. Work through these steps to introduce the concept to your child.

1. **Tell your child that to add unlike fractions, you must convert them to fractions with the same denominator using the same technique they used to compare fractions:**

   $$\frac{1}{3} + \frac{1}{2} =$$

   $$\frac{1}{3} \times \frac{2}{2} = \frac{2}{6}$$

   $$\frac{1}{2} \times \frac{3}{3} = \frac{3}{6}$$

2. **After converting the fractions so they have the same denominators, your child can simply add the numerators, like so:**

   $$\frac{2}{6} + \frac{3}{6} = \frac{5}{6}$$

3. **Repeat this process for the following unlike fractions:**

   $$\frac{1}{4} + \frac{1}{5} = \qquad\qquad \frac{1}{4} \times \frac{5}{5} = \frac{5}{20}$$

   $$\frac{1}{5} \times \frac{4}{4} = \frac{4}{20} \qquad\qquad \frac{5}{20} + \frac{4}{20} = \frac{9}{20}$$

Use Worksheet 25-2 at www.dummies.com/go/teachingyourkidsnewmathfd to practice adding unlike fractions. Help your child with the first few, and then ask them to try to complete the rest of the worksheet.

Subtracting Unlike Fractions

Unlike fractions aren't fractions you dislike. They're simply fractions that have different denominators (you know, the bottom number). Fortunately, the process for subtracting unlike fractions is similar to the process your child used to add them. Try working through these examples together:

1. Tell your child, "Often, you must subtract unlike fractions, which means you're working with fractions that have different denominators, like these."

$$\frac{1}{2} - \frac{1}{4} =$$

2. Explain to your child that to subtract the fractions, they must first convert them to fractions with the same denominator using the same process they used for addition, as shown here:

$$\frac{1}{2} - \frac{1}{4} =$$

$$\frac{1}{2} \times \frac{4}{4} = \frac{4}{8}$$

$$\frac{4}{8} - \frac{2}{8} = \frac{2}{8}$$

$$\frac{1}{4} \times \frac{2}{2} = \frac{2}{8}$$

Download and print Worksheet 25-3 from www.dummies.com/go/teachingyour kidsnewmathfd so that your child can do some more practice with subtracting unlike fractions. Help your child with the first few, and then encourage them to complete the rest of the worksheet on their own.

Multiplying Fractions

Say to your child, "You have learned to add and subtract fractions. Now, you will learn how to multiply fractions." The good news is that multiplying fractions is actually easier that adding and subtracting fractions because you don't have to worry about first creating like fractions that have the same denominator. Instead, you can simply multiply the fractions.

Tell your child, "To multiply two fractions, you multiply the two numerators and then the two denominators."

$$\frac{1}{2} \times \frac{3}{4} = \frac{3}{8}$$

Repeat this process for the following expression:

$$\frac{2}{3} \times \frac{4}{5} =$$

In this case, you should get.

$$\frac{2}{3} \times \frac{4}{5} = \frac{8}{15}$$

Worksheet 25-4 at www.dummies.com/go/teachingyourkidsnewmathfd offers practice for multiplying fractions. Help your child with the first few problems, and then ask them to complete the worksheet.

Dividing Fractions

Division operations make working with fractions extra interesting. The process of dividing fractions is very similar to that of multiplying fractions, with the difference being that we flip the second fraction, creating its reciprocal, and then we multiply. The reciprocal of ⅔, for example, is ³⁄₂. Likewise, the reciprocal for ¾ is ⁴⁄₃.

Say to your child, "To divide two fractions, you flip the second fraction (using the reciprocal) and then multiply the resulting fraction."

$$\frac{1}{3} \div \frac{4}{5} =$$

$$\frac{1}{3} \times \frac{5}{4} = \frac{5}{12}$$

Repeat the process for the following expression:

$$\frac{2}{3} \div \frac{3}{4} =$$

You should get this result:

$$\frac{2}{3} \div \frac{3}{4} =$$
$$\frac{2}{3} \times \frac{4}{3} = \frac{8}{9}$$

Worksheet 25-5 contains a worksheet for dividing fractions. Help your child with the first few, and then ask them to complete the worksheet.

Working with Mixed Numbers

Your child has just learned to perform math operations with fractions that don't include a whole number. In this section, they will learn to perform the same operations with mixed numbers that have a whole number and fractional part, such as 1½. To do so, your child will first convert the mixed numbers to improper fractions (fractions such as ⁵⁄₄ in which the numerator is bigger than the denominator). Then, they can perform the operation.

1. **Present the following expression to your child:**

$$1\frac{1}{2} + 2\frac{2}{3} =$$

2. **Tell your child that they need to start by first converting the mixed numbers to improper fractions.**

You do this by multiplying the whole number by its corresponding denominator and adding that number to the fraction's numerator:

$$1\frac{1}{2} + 2\frac{2}{3} =$$
$$\frac{3}{2} + \frac{8}{3} =$$

3. **When you are adding or subtracting, make sure that the fractions have a common denominator.**

If they don't, convert them to fractions with common denominators, as shown here:

$$\frac{3}{2} + \frac{8}{3} =$$

$$\frac{3}{2} \times \frac{3}{3} = \frac{9}{6} \qquad \frac{8}{3} \times \frac{2}{2} = \frac{16}{6}$$

4. **Add the results:**

$$\frac{9}{6} + \frac{16}{6} = \frac{25}{6}$$

5. **Convert the improper fraction to a mixed number as discussed in Chapter 18.**

$\frac{25}{6}$ is an improper fraction, so it needs to be converted, like so:

$$4\frac{1}{6}$$

Subtracting mixed numbers uses the same process. Here's an example to work through with your child.

1. **Consider the following expression:**

$$2\frac{2}{3} - 1\frac{1}{2} =$$

2. **Convert the mixed numbers to fractions:**

$$\frac{8}{3} - \frac{3}{2} =$$

3. **Find a common denominator:**

$$\frac{8}{3} \times \frac{2}{2} = \frac{16}{6} \qquad \frac{3}{2} \times \frac{3}{3} = \frac{9}{6}$$

4. **Perform the subtraction:**

$$\frac{16}{6} - \frac{9}{6} = \frac{7}{6}$$

5. **Convert the result to a mixed number as discussed in Chapter 18:**

$$1\frac{1}{6}$$

Worksheet 25-6 at www.dummies.com/go/teachingyourkidsnewmathfd includes expressions for adding and subtracting mixed numbers. Help your child with the first few, and then ask them to complete the worksheet.

Multiplying mixed numbers

As previously discussed, mixed numbers have a whole part and a fractional part, such as 1½. Sometimes you need to multiply them — not that often in real life, but it is a skill that's taught in math classes. In this lesson, you will teach your child how to work through these problems to help them be successful with school assignments.

1. **To start, you must convert the mixed numbers to improper fractions by multiplying each whole number by its corresponding denominator and then adding that result to the fractional part:**

$$2\frac{2}{3} \times 1\frac{1}{2} =$$
$$\frac{8}{3} \times \frac{3}{2} =$$

2. **Multiply the two fractions:**

$$\frac{8}{3} \times \frac{3}{2} = \frac{24}{6}$$

3. **Convert the improper fraction:**

4

Dividing mixed numbers

Just as there are times when you must multiply mixed numbers, there are times when you must divide them. You remember that time — when you took the SAT or ACT test. Dividing mixed numbers doesn't come up too often outside those standardized tests, but it's a good skill to have anyway. In this lesson, you will teach your child how to divide mixed numbers.

1. **Convert the mixed numbers to improper fractions by multiplying each whole number by its corresponding numerator and adding that result to the fractional part:**

$$3\frac{1}{3} \div 1\frac{1}{2} =$$
$$\frac{10}{3} \div \frac{3}{2} =$$

2. Flip the second fraction (creating the reciprocal) and multiply:

$$\frac{10}{3} \times \frac{2}{3} = \frac{20}{9}$$

3. Convert the improper fraction to a mixed number:

$$2\frac{2}{9}$$

Whew! This chapter has covered a lot of skills. There's just one more worksheet that will round out the practice of multiplying and dividing mixed numbers: Worksheet 25-7 (found at www.dummies.com/go/teachingyourkidsnewmathfd). Help your child with the first few problems, and then ask them to complete the worksheet.

Chapter **26**

Knowing the Point of Decimals

Let's define decimals. Decimals are numbers that have a decimal point, such as 5.95. There, that was easy!

Solving problems with decimal points takes a little more work, though. You may be tempted to dust off your old calculator or fire up your phone's calculator app to solve problems this chapter presents. Don't do it! You can work with decimals without a calculator and so can your child.

I start with money to introduce the concept — which is far more fun than my fraction example with ½ cup of flour to make grandma's gravy. Because your child will work with money and decimal numbers throughout their life, this chapter's concepts are very important.

TIP

Before you get started, you'll want to have the following supplies on hand:

» Spare change (quarters, nickels, and dimes)

» A few sheets of scrap paper.

Revisiting Money

All math students at some point ask, "When would I use this in real life?" Well, everyone deals with money, and money has decimal points, so money is a good place to start for this chapter.

You've worked through several chapters since you and your child talked about money, so follow these steps to get a refresher and introduce decimals to them:

1. **Place one quarters, two dimes one nickel, and three pennies in front of your child, and ask them to count the coins and tell you how much they have.**

2. **Explain that you write amounts of money by starting with a dollar sign ($), writing the dollar amount, following with a period, and writing the change.**

   For the change your child currently has, use a piece of paper to write the following amount:

   $0.53

   Explain that this amount states that you have 0 dollars and 53 cents.

3. **Ask your child to count the amount of money in three quarters, two dimes, four nickels, and two pennies.**

4. **Write the amount on a piece of paper, like so:**

   $1.17

5. **Ask your child to read the number out loud.**

6. **Hand your child $1.28 in change, and ask them to count the money and write the amount on a piece of paper.**

   You may need to remind them of the order for writing the amount: the dollar sign, the number of dollars, a decimal point, and the change.

Use Worksheet 26-1 from www.dummies.com/go/teachingyourkidsnewmathfd to help your child practice counting change and writing the amount in the correct format.

Knowing the Tenths and Hundredths Places

After revisiting counting money and writing dollar amounts, which is a somewhat familiar concept, you can begin talking decimals in a more general sense.

Recognizing the tenths place

1. Say to your child, "When we count out money and other numbers, there will be times when we have a part of a whole number, such as the cents that make up one dollar. In such cases, we use a period, which is known as a decimal point, as shown in these numbers."

   5.85   3.34   7.23   8.48

   Explain to your child that you read such numbers as "five point eight five," "three point three four," and so on.

2. Remind your child that they know the ones, tens, hundreds, and thousands places, and ask them to write the digits in the following number in the correct placeholders:

   5237

3. Show your child the following number and ask them to write the numbers in the corresponding places, noting the location of the decimal point:

   534.56

			.		

   Their answer should look like this:

5	3	4	.	5	6

4. Have your child read the number out loud and then repeat this process for the following number:

132.77

			.		

5. Explain to your child that they can think of the numbers in the tenths place as being the same as dimes.

It takes 10 dimes to create 1 dollar. Consider the following number:

273.3

6. Tell your child that they can represent the number visually as shown here:

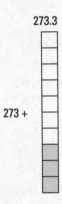

7. Show your child the following number and ask them to color in the number of tenths on the following visualization:

17.5

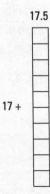

8. Have your child color in the correct number of boxes for the tenths places in the following numbers:

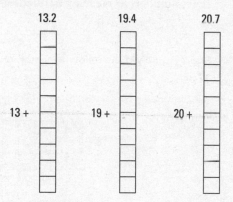

9. Present the following illustration to your child and ask them to write the corresponding number of tenths in the appropriate boxes:

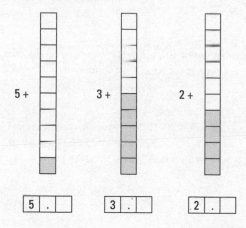

TIP

Use dimes to help your child understand the concepts of the tenths place, reminding them that it takes 10 dimes to make a dollar. If they have 3 dimes, they have 0.3.

Recognizing the hundredths place

Sometimes, decimal numbers don't stop at the tenths place. A decimal such as 1.25 goes one step beyond to the hundredths place. With the following steps, your child will understand and identify the hundredths place within decimal numbers:

1. **Present the following number to your child and explain, "The following number has two digits to the right of the decimal point. You can think of these numbers as meaning 22 hundredths."**

   17.22

2. **Show your child the following visual representation of the hundredths in 17.22:**

   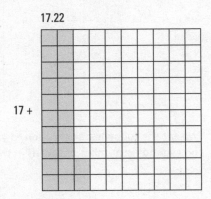

   17.22

   17 +

3. **Present the following number to your child and look at the visual representation that follows it:**

   3.33

   TIP

   Use pennies to help your child understand the concept of the hundredths place, reminding your child that it takes 100 pennies to make a dollar. If they have 53 pennies, they have 0.53.

   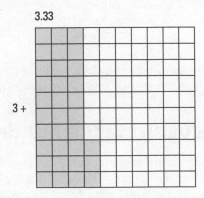

   3.33

   3 +

4. Present the following illustration to your child and ask them to write the corresponding number:

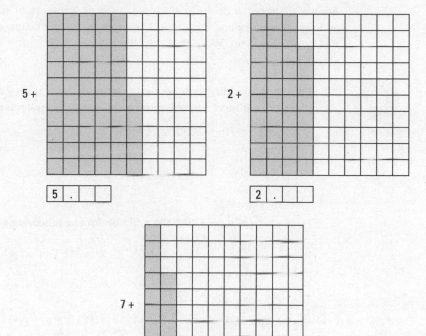

5 +

| 5 | . | | |

2 +

| 2 | . | | |

7 +

| 7 | . | | |

5. Present the number 1.33 to your child and have your child shade the appropriate number of boxes within the following illustration:

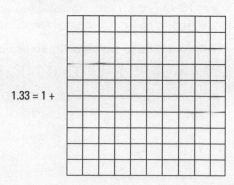

1.33 = 1 +

Adding Numbers with Decimal Points

Adding numbers with decimal points, such as amounts of money, is not unlike adding other numbers, but there is the special consideration of keeping the decimal points lined up. Work through this process with your child in the following steps:

TIP

1. **Explain that adding numbers with decimal points is very similar to standard addition, but you must include the decimal point in your result.**

   Show your child this example:

   ```
     3.12
   +1.23
     4.35
   ```

2. **Have your child complete the addition for the following expression:**

   ```
     2.27
   + 3.42
   ```

   They should get this result:

   ```
     2.27
   + 3.42
     5.69
   ```

3. **Present the following expression to your child:**

   ```
     1.59
   + 1.23
   ```

   In this case, when your child adds the numbers with decimal points, they will need to carry the result of the addition over to the next number. Explain that they should perform such carry operations as they normally would.

4. **Ask your child to complete the equation.**

   They should get this answer:

   ```
        1
     1.59
   + 1.23
     2.82
   ```

Subtracting Numbers with Decimal Points

Every time you hand someone cash to pay for something at a store, the cashier is subtracting numbers with a decimal point to make your change. In this section, your child will learn to do the same. Your child will use the subtraction skills they already possess and align numbers on the decimal point so they can later write the decimal point in their result.

1. **Explain to your child that subtracting numbers with decimal points follows the same process as normal subtraction, with the exception that they must remember to include the decimal point in the answer, as shown here:**

    $$\begin{array}{r} 3.42 \\ -2.21 \\ \hline 1.22 \end{array}$$

2. **Also explain that as they subtract numbers with decimal points, there may be times when they must borrow from the next column before they can subtract.**

    Tell your child that the carry process will not change. Consider the following expression:

    $$\begin{array}{r} 3.17 \\ -1.78 \end{array}$$

    Have your child complete the expression. They should get this result:

    $$\begin{array}{r} 2\ 10 \\ \cancel{3}.\cancel{1}^{1}7 \\ -1.7\ 8 \\ \hline 1.3\ 9 \end{array}$$

Worksheet 26-2 at www.dummies.com/go/teachingyourkidsnewmathfd includes both addition and subtraction problems to practice working with decimal points. Help your child work through the first few problems, and then ask them to complete the worksheet.

Multiplying Numbers with Decimal Points

Just as your child must add and subtract numbers with decimals, there will be times when they must multiply them, such as to understand why it's so expensive to put 20.1 gallons of gas in a car when the price is $3.99 a gallon.

1. **Explain to your child that multiplying numbers with decimal points is much the same as normal multiplication; the only difference is in determining where to place the decimal point within the result.**

   Show your child the following expression and ask them to complete the multiplication as they normally would (without regard for the decimal points). They should get the following result:

   $$\begin{array}{r} 3.1 \\ \times\, 2.5 \\ \hline 155 \\ 620 \\ \hline 775 \end{array}$$

2. **Tell your child that they determine where to put the decimal point in the result by looking at each number they multiplied and counting the number of digits that appear to the right of the decimal point.**

   In this example, there are a total of two numbers to the right of the decimal point (one in the top number and one in the bottom number). Consequently, they should place the decimal point to the left of the two digits in the rightmost positions.

   Adding a decimal point to the result of the multiplication in Step 1 looks like this:

   7.75

3. **Ask your child to try multiplying the following expression:**

   $$\begin{array}{r} 5.2 \\ \times\, 3.1 \\ \hline \end{array}$$

They should get this result:

$$5.2$$
$$\times 3.1$$
$$\underline{52}$$
$$\underline{1560}$$
$$16.12$$

Worksheet 26-3 at www.dummies.com/go/teachingyourkidsnewmathfd is filled with examples of problems for multiplying numbers with decimal points. Help your child with the first few, reminding them where to place the decimal point, and then ask them to complete the worksheet.

Recognizing Equivalent Decimals

When your child works with decimal numbers, there will be many times when they encounter numbers such as 7.5 and 7.50 or 8.3 and 8.30. In this section, your child will learn that such numbers are equal — meaning the zero at the end of the number does not change the value.

Present the following numbers to your child:

1.1   1.10

1. **Remind your child that they know how to use boxes to represent the numbers, as shown here:**

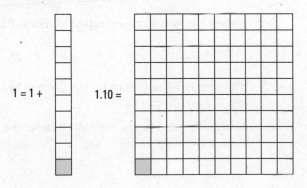

$$1 = 1 + \qquad 1.10 =$$

Explain to your child that the illustrations have the same number of shaded boxes, which means that 1.1 and 1.10 are equal.

2. **Present the numbers 3.5 and 3.50 to your child and have them shade the boxes shown here:**

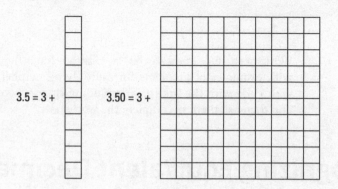

$3.5 = 3 +$ $3.50 = 3 +$

3. **Ask your child if 3.5 and 3.50 are the same. If they say no or are unsure, repeat the process with numbers such as 2.2 and 2.20.**

Dividing Numbers with Decimal Points

If long division wasn't fun enough for you, now it's time to throw decimal points into the festivities. The good news is that the division process doesn't change, although the decimal points may make it look more complicated.

1. **Present the following expression to your child:**

$$25\overline{)755.5}$$

Help your child work through the division by asking, "Can you divide 25 into 7? How about into 75?"

2. **Because 75 ÷ 25 = 3, you write the 3 above the division bar and subtract 75 as shown here:**

$$\begin{array}{r} 3 \\ 25\overline{)755.5} \\ \underline{75} \\ 0 \end{array}$$

3. **Bring down the 5. Because 25 cannot divide into 5, write a 0 on the top bar and bring down the last 5.**

$$
\begin{array}{r}
30 \\
25\overline{)755.5} \\
\underline{75} \\
055
\end{array}
$$

Note that the 0 you just wrote is above the number that's next to the decimal point in 755.5.

4. **Tell your child that the decimal point in the answer goes directly above where it is in the original number.**

Have your child bring up the decimal point to the top row, placing it to the right of the 30, like so:

$$
\begin{array}{r}
30. \\
25\overline{)755.5} \\
\underline{75} \\
055
\end{array}
$$

5. **Because they can divide 25 into 55, have your child write down the 2 and subtract 50, as shown here:**

$$
\begin{array}{r}
30.2 \\
25\overline{)755.5} \\
\underline{75} \\
055 \\
\underline{50} \\
5
\end{array}
$$

6. **Tell your child that they can think of the numbers 755.5 and 755.50 or 755.500 as the same, so they can bring down zeros as often as they need to complete the division problem.**

In this case, they can divide 25 into 50, so write the 2 at the far right of the number on the top bar and subtract the 50, like this:

$$
\begin{array}{r}
30.22 \\
25\overline{)755.5} \\
\underline{75} \\
055 \\
\underline{50} \\
50 \\
\underline{50} \\
0
\end{array}
$$

7. **Repeat this process for the following expression:**

$$11\overline{)137.5}$$

The result you and your child get should look like this:

```
        12.5
   11)137.5
      11
      27
      22
       55
       55
        0
```

Worksheet 26-4 at www.dummies.com/go/teachingyourkidsnewmathfd includes division problems with decimal points. Help your child as needed to complete the worksheet.

Chapter **27**

Looking at Lines, Angles, Areas, and Perimeters

G eometry is the study of points, lines, angles, surfaces, and shapes. In this chapter, your child will start using geometry! If you've ever measured a room or bought a carpet, you've used geometry, and you'll bring real-world knowledge to your teaching.

By the end of this chapter, your child will know how a shape's perimeter differs from its area and how to calculate each. You will also introduce them to the symbol for pi (π), which they'll use when working with circles. As an added bonus, you may find yourself starting to laugh at the math jokes on reruns of the TV show *The Big Bang Theory*.

Recognizing Parallel and Perpendicular Lines

Throughout this book, you and your child have used number lines to count and subtract numbers. Here's one of your old friends:

7 + 5 =

Tell your child, that they can think of a line as a figure that goes on forever. In the case of a number line, they can also add one more to the count shown, so a number line does not have to stop where the labels stop at one million, two million, or even three million. Instead, it can just go on.

Lines don't always appear solo, though. Talk with your child about the different ways lines work together.

Consider the following lines and say to your child, "These lines *intersect* or cross one another."

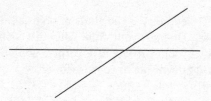

Look at these lines with your child:

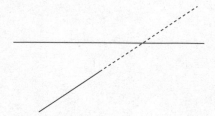

In this case, the two lines do not yet intersect. However, if you extended them, they would, as indicated by the dotted line.

There are times when two lines, even if you extend them, will never intersect. Such lines are called *parallel lines*, and they look like this:

Explain to your child that they will see parallel lines in shapes such as the square or rectangle shown here. The top and bottom lines are parallel to each other and the two side lines are parallel to one another:

Recognizing Common Angles

When two lines intersect, they form angles. The angle the lines create depends on the direction of each line. We describe angles in degrees. A rectangle, for example, has four 90-degree angles. Knowing that, I still don't understand what it means when someone says, "he must have an angle," or "what's his angle?" Ulterior motives should have nothing to do with math!

You can define an *angle* for your child as being two lines that intersect, like these examples:

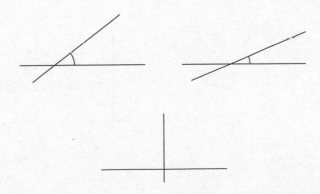

Explain that they can use a tool, called a *protractor*, to determine specifics about angles. Figure 27-1 shows a protractor.

You can measure many things using a tape measure or ruler using feet and inches as the units of measurement. When you measure angles with a protractor, you specify the measurement in *degrees*. The bigger the angle, the greater the number of degrees. Figure 27-2 shows several common angles and their corresponding measurements in degrees.

FIGURE 27-2:
Common angles and their degrees.

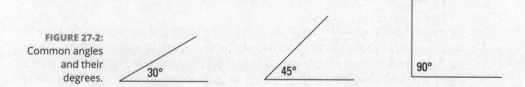

Ask your child to pay special attention to the 90-degree angle. Explain that often, 90-degree angles are indicated in a drawing with a small square on the angle, as shown here:

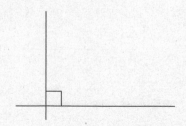

Further explain that each corner of a square or rectangle forms a 90-degree angle:

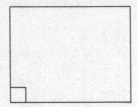

REMEMBER

You may need to remind your child that parallel lines are lines that never intersect. In contrast, perpendicular lines are lines that intersect to form a 90-degree angle, as shown here:

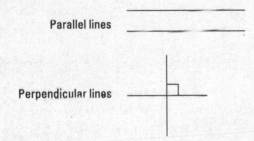

Parallel lines

Perpendicular lines

Calculating the Perimeter of a Rectangle

When drawing representations of real things — such as rooms, your backyard, or a building — people often use a rectangle. For example, the following rectangle could represent a bedroom, and the measurements correspond to the length of each wall:

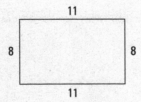

Say to your child, "Often we will want to know the total distance around the rectangle, which is called the *perimeter*. To calculate the perimeter, you add up the measurement for each side of the rectangle, as shown here."

$$
\begin{array}{r}
11 \\
+\ 8 \\
+11 \\
+\ 8 \\
\hline
38
\end{array}
$$

Present the following rectangle to your child and ask them to calculate the rectangle's perimeter by adding the sides:

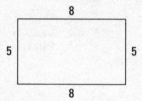

Your child should get the following total:

$$
\begin{array}{r}
8 \\
+\ 5 \\
+\ 8 \\
+\ 5 \\
\hline
26
\end{array}
$$

Understanding the Special Rectangle: The Square

Present the following square to your child:

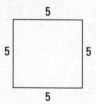

Tell your child that a square is a special type of rectangle for which all the sides are the same. Say, "To compute the perimeter of a square, you can add the measures for each side, just as you did for a rectangle, like this."

```
   5
 + 5
 + 5
 + 5
  20
```

Then explain that because the length of each side of the square is the same, you can also calculate the perimeter of a square by multiplying 4 times the length of one side. Here's an example with a different square:

```
   6
 + 6        6
 + 6      × 4
 + 6       24
  24
```

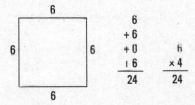

Ask your child to calculate the perimeter of the following square:

Calculating the Perimeter of a Triangle

Perimeter is perimeter, whether you're talking about the distance around a rectangle, a square, or a triangle. Use these steps to practice calculating the perimeter of triangles:

1. **Say to your child, "You have learned to calculate the perimeter of a square and a rectangle. In a similar way, to calculate the perimeter of a triangle, you again add the measurement of each side."**

Present the following triangle to your child:

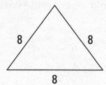

2. **Ask your child to calculate the triangle's perimeter.**

They should get the following result:

$$\begin{array}{r} 8 \\ + \ 8 \\ + \ 8 \\ \hline 24 \end{array}$$

Worksheet 27-1 at www.dummies.com/go/teachingyourkidsnewmathfd is filled with squares, rectangles, and triangles for which your child can calculate the perimeter. Supervise your child as they do the first few, and then let them go it alone to complete the rest of the worksheet.

Knowing the Parts of a Circle

You see examples of circles every day in the real world — such as wheels, glasses, balls, and so on. So a circle isn't a foreign concept. The names of the parts of a circle may be somewhat unfamiliar, though. Use the following steps to introduce your child to the parts of a circle:

1. **Tell your child that you often need to know the size of a circle, and to figure out that measurement, you must measure a line, called the *diameter*, that cuts the circle in half.**

Here's a circle with the diameter drawn on it:

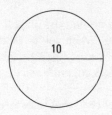

In this case, the diameter of the circle is 10.

2. **Ask your child to identify the diameter of the following circles:**

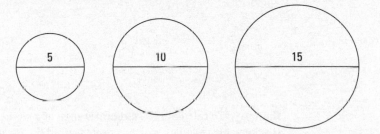

3. **Explain to your child that to perform many mathematical operations on a circle, you use one-half of the diameter, which is called the *radius*.**

The following circle has a radius of 5:

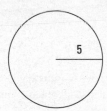

Ask your child to identify the radius of the following circles:

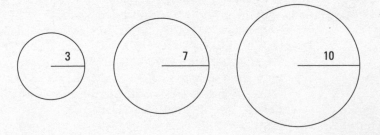

Calculating the Circumference of a Circle

The distance around the outside of a square, a rectangle, and a triangle is called the perimeter. When you're talking about the distance around a circle, you use a different word: *circumference*. Use these steps to explain to your child how to calculate a circle's circumference:

1. **Tell your child that to determine the circumference, you must know the circle's radius.**

The radius is marked on this circle:

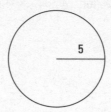

2. **Then say, "To calculate the circumference of a circle, you also use a special value called pi, which is represented by the symbol π."**

$$\pi = 3.14$$

3. **Explain that the following equation calculates the circumference of a circle:**

$$2 \times \text{radius} \times \pi$$
$$2 \times 5 \times \pi$$
$$2 \times 5 \times 3.14 = 31.4$$

TIP

You may have to remind your child that they need to work from left to right when multiplying multiple numbers.

4. **Help your child determine the circumference of the following circle:**

They should get the following result:

$$2 \times \text{radius} \times \pi$$
$$2 \times 3 \times \pi$$
$$2 \times 3 \times 3.14 = 18.84$$

Worksheet 27-2 at www.dummies.com/go/teachingyourkidsnewmathfd has circles galore so your child can practice calculating circumferences to their heart's content. The math here is a little trickier than on some worksheets, so help your child as much as necessary to complete the worksheet.

Calculating the Area of a Shape

Perimeter and circumference aren't the only helpful measurements of shapes. Sometimes, you need to know the amount of space inside of a shape, which is called the shape's *area*. The shading on the following shapes represents the area:

Calculating the area of a rectangle or square

Painting a room is one real-life task where calculating area comes in handy. You need to know the area of the wall so that you can buy enough paint to cover it. Use these steps to illustrate how to calculate the area of a rectangle or square:

1. **Explain that to calculate the area of a square or rectangle, like the one shown here, your child will multiply the height of the rectangle times its width:**

10

8

Ask your child to identify the width and the height of the rectangle and then do the multiplication to calculate the area. They should get this result:

$10 \times 8 = 80$

2. **Present the following shapes to your child:**

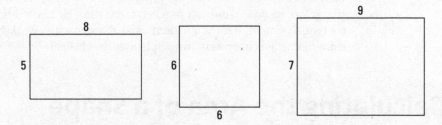

Help your child calculate each shape's area. You should get:

$$\begin{array}{r} 8 \\ \times\ 5 \\ \hline 40 \end{array}$$

$$\begin{array}{r} 6 \\ \times\ 6 \\ \hline 36 \end{array}$$

$$\begin{array}{r} 9 \\ \times\ 7 \\ \hline 63 \end{array}$$

Calculating the area of a triangle

Say to your child, "Just as there are times when you must calculate the area of a square or rectangle, you may also need to calculate the area inside of a triangle."

Explain that to calculate the area of a triangle, you must know the triangle's base (the length of the triangle's bottom line) and height, as shown on this triangle:

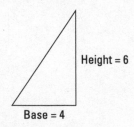

Use these steps to work through an example of calculating the area of a triangle.

1. **Tell your child that to calculate the area of a triangle, you multiply ½ times the triangle's base times the triangle's height; the equation looks like this:**

$$\text{Area} = \frac{1}{2} \times \text{base} \times \text{height}$$

2. **Present the following triangles to your child and help them calculate each triangle's area:**

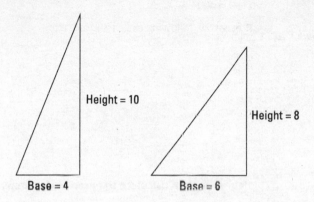

Height = 10

Base = 4

Height = 8

Base = 6

You should get the following results:

$$\frac{1}{2} \times 4 \times 10 =$$

$$2 \times 10 =$$

$$20$$

$$\frac{1}{2} \times 6 \times 8 =$$

$$3 \times 8 =$$

$$24$$

Worksheet 27-3 at www.dummies.com/go/teachingyourkidsnewmathfd includes a variety of squares, rectangles, and triangles with measurements marked for practicing how to calculate area. Help your child as much as necessary for them to complete the worksheet.

Calculating the area of a circle

Calculating the area of a circle is a higher-level skill normally taught in sixth or seventh grade. However, because your child has learned the symbol pi, they will understand the equation to calculate a circle's area, as I discuss in this section.

REMEMBER

1. **Explain to your child that to calculate the area of a circle, they will use the following expression:**

$$\text{Area} = \text{radius} \times \text{radius} \times \pi$$

Pi (π) is equal to 3.14.

Present the following circle to your child:

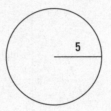

2. **Help your child calculate the area of the circle.**

The result should be as follows:

$$\text{Area} = \text{radius} \times \text{radius} \times \pi$$
$$= 5 \times 5 \times \pi$$
$$= 5 \times 5 \times 3.14$$
$$= 78.5$$

7
The Part of Tens

Ten suggestions for helping you teach your child math

Ten ways to round out your child's education

Chapter **28**

Top Ten Things to Remember about Teaching Your Child Math

Teaching your child math, especially new math, may seem daunting at first. Not only must you remember math that you may not have used for years, but there are also new approaches that you will have to learn before you can teach them to your child.

I commend for taking the steps and for setting aside the time to teach your child math. As you get started in the teaching process, keep the following ten things in mind.

Establish Your Child's Baseline

Most children who struggle with a math topic normally do so because they did not develop a strong foundation with respect to a previous math topic. This book starts with kindergarten math, and each chapter contains the key concepts that the people who define math standards deem important. You should take time to review each grade level's concepts with your child.

I've included accompanying worksheets for the key concepts in many chapters. You can find them at www.dummies.com/go/teachingyourkidsnewmathfd. Before you start teaching math at your child's current grade level, you should download and print the worksheets for the previous grade level to test your child's core knowledge. If a topic in a previous grade seems new to your child, stop and review the corresponding topic until your child masters it. Although your child may be in fifth grade, they may not know all the skills presented in earlier grades. Use the chapter worksheets to establish a baseline for your child's current skill set.

Set Aside 15 Minutes a Day, Every Day

Mastering math skills requires practice and repetition. To ensure your child's success, you should plan to set aside 15 minutes during an unstressful part of each day to review previous concepts and introduce new ones. Try to establish a regular schedule that you can meet each day, such as 15 minutes before or after dinner. By scheduling at a consistent time, you and your child can plan and prepare for the interaction.

Keep the Focus on Your Child

Find a quiet place that is free from distractions such as the TV, where you and your child can focus on the math concepts and exercises. You and your child will find greater success if you focus entirely on the task at hand. Try to set aside your phone and laptop for 15 minutes so that you can focus on your child as your child focuses on the work. Your child will appreciate your commitment to their success.

Be Patient

Children learn at different rates. Learning math is not a race, so if your child requires additional time to master a concept, that's fine. Remember, the math concepts that this book presents will be new to your child. Your goal is to lay a solid foundation for your child's long-term success.

Use Additional Resources

New math is, well, new. The good news is that there are many resources on the web that include videos and additional practice worksheets that can help you. This book's companion website at www.dummies.com/go/teachingyourkidsnewmathfd contains many additional practice worksheets. Take advantage of them.

Take Time to Master Each Grade Level

If your child has trouble with a new topic, don't be afraid to take a step back to a previous grade level. Math can be challenging. If your child becomes frustrated, take time to review concepts they have previously mastered. Your child will regain their confidence while improving their skills.

Point Out Math in the Real World

The impact of math extends far beyond books. You can find countless examples of math in the real world:

>> Finding TV shows on numbered channels

>> Keeping score in sporting events

>> Tracking prices at the grocery store

>> Using fractions while cooking

>> Observing charts and graphs on a business channel

By pointing out math in the world around your child, you will reduce questions such as, "Why do I need to know this?"

Praise the Things Your Child Does Well

Remember to keep a positive attitude. Praise the things your child does well, ideally in front of others. When your child makes a math error, and they will make many, keep the focus on the problem — meaning, the answer is wrong, not your child. Avoid expressions such as, "You got this one wrong." Instead, use: "This answer is not correct. Let's find out why." Remember, the baseball player Ted Williams, who was the greatest batter of all time, didn't get a hit at six out of ten tries. People who try will make mistakes. Build on successes.

Talk with Your Child's Teacher

Your child spends almost eight hours every day with their teacher. The teacher knows your child and their skills well. Let them know that you are working on math with your child. The teacher may have additional resources or insights that will help you. Partner with the teacher on your child's success.

Keep It Fun

Math can be fun. The better you are at it, the more fun it will be. Keep the 15 minutes that you dedicate to your child's math success a positive and fun experience. You and your child should enjoy the learning process. You might search the web with your child for songs and videos that relate to the topic at hand. Your child will enjoy the time you spend together searching.

Chapter **29**

Ten Next Steps in Your Child's Education

Math is important. Your child will use math every day for the rest of their life, so it's great that you are helping them establish a solid math foundation. That said, there is more to life than math! To round out your child's other skills, you should consider the following ten next steps in your child's education.

Read with Your Child

Reading is just as important as math, as your child will also use reading skills every day of their life. To help your child succeed, you should spend time reading with them. For the first few years, you will teach your child to read. Try to find time for your child to read to you each day. Let your child see you reading, and ideally, set aside time when you can read together.

Be Involved

Learning requires participation. Ask your child (and their teacher) what topics they are learning. You likely know something about each topic, and your child will enjoy it when you share what you know. For example, each night at dinner you might ask your child, "What did you learn today?" *Nothing* is never a good answer! If that's your child's response, then say, "Well, I guess you will learn something tonight. Would you like to do math or reading?" You will probably only have to say this one time before your child finds a better answer in the future. Partner with your child's teacher. They have the most insight into how your child is doing in school. Ask them what your child is learning, as well as what they *will be* learning. With such knowledge in hand, you can better prepare your child.

Keep Video Games to a Reasonable Amount

Kids love video games, and allowing your child to play video games sometimes is okay. While playing games, your child will learn how to use technology such as interacting with a device, how to start programs, run apps, and more. Video games can also teach your child tactical and strategic thinking. Multiuser games can teach your child to collaborate with others. They can also be an opportunity for you to bond with your child. Take an interest in the games that your child is playing, and have them explain the game to you and then find time to watch and support your child. These are some of the good aspects to video games.

However, the amount of time your child plays such games must be reasonable. Establish limits and then stick to them. Let your child know that they can play a game until 3:30, for example, and then the game is off for the day.

Help Your Child Learn Computer Skills

Computers and math are closely related. Like math, people either love computers or they hate them. Like math, most of us must use computers every day. You need to develop your child's computer skills such as typing at a keyboard. There are many good typing games that your child will enjoy as they build this critical skill. There are also many great math applications and games available for computers and phones. Find them and help your child get started.

Implement a Word of the Day

Kids who read regularly develop a better vocabulary than those who don't. Kids with larger vocabularies tend to do better in school. Implement a word-of-the-day program at home for which you teach your child (and maybe yourself) a new word each day. To start, you can introduce and define the new words. Later, you can give your child a dictionary and allow them to present the new word each day.

Ensure Adequate Sleep Time

Your child should get at least eight hours of sleep each night. Children who are rested do better at school. Establish a consistent bedtime for your child. This may require that you set evening schedules that do not encourage your child to stay up late.

Encourage Learning a Musical Instrument

Children will find learning a musical instrument fun. Beyond that, learning a musical instrument will improve your child's math skills, and research has shown that learning music develops different parts of the brain. Practicing an instrument requires patience and persistence, and your child's success with music will increase their confidence. Finally, the ability to play an instrument will increase social opportunities for your child.

Learn a Second Language Together

Research shows that children who learn a second language early in life will develop parts of their brain that would otherwise not develop. If you don't know a second language, relax. There are many great apps you can use to get your child started. You might even decide to learn the language together. Your child would love to use a different language to communicate with you.

Listen and Talk to Your Child

Being a parent is hard. That said, being a child can be hard, too. There are many things to learn, struggles to fit in socially, and much more. Take time to talk with your child and, more importantly, listen to the things they say. Your child should know that they can always come to you with problems and questions.

Involve Your Child

Often, the best way to learn is through life experiences. As such, you should involve your child in the things that you do. If you are buying a new car or a new house, have discussions in which your child can participate. Doing so will teach your child to listen, ask appropriate questions, and feel included. Children of entrepreneurs tend to become entrepreneurs. That's often because they have watched, learned, and participated in conversations about business. The point is, the more you can involve your child, the more they will learn.

Index

A

addition
- of 1 minute to times with 50 minutes, 100–101
- of 10 mentally, 145
- of 100 mentally, 146–147
- creating equivalent expressions using, 160–161, 162–163
- of factions, 274
- flash cards through 10, 95–96
- identifying
  - tens and ones places, 82–85
  - tens and ones places using worksheets, 84–85
  - tens on number lines, 83–84
- of large numbers
  - using boxes, 88–89, 141–142
  - using number lines, 86–88
  - using open number lines, 176–177
  - without regrouping, 140–144
  - without using boxes, 142
- of like fractions, 226–227
- in mental math, 133–138
- mixing with subtraction problems on same worksheet, 95
- of multi-digit numbers
  - through 1,000 (old math), 254–255
  - through 10,000, 255–256
  - using number lines (new math), 256–258
  - using rounding, 258–260
- of numbers
  - with decimal points, 310
  - through 20 mentally, 134–136
  - through 100 without boxes, 89
  - through 100 without regrouping, 85–86, 89
- operations and, 237–238
- practicing using index cards, 62–63
- with regrouping, 165–180, 252
- repeated, 186–187
- subtraction and, 73
- of three rows of multi-digit numbers, 252–254
- through 20, 105–106
- of unlike fractions, 295–296
- using boxes and regrouping for, 166–169
- using decomposition, 171–172
- using flash cards, 66–67
- using new math for large numbers, 176–179
- using number lines, 65–66
- using regrouping without boxes for, 169–170
- using straws for, 62
- using worksheets, 68
- without regrouping, 166
- on worksheets, 64, 65
- of zero to numbers, 74, 75–76

address, learning your, 28–29

analog clocks, 102–105

angles, common, 319–321

area, calculating for shapes, 327–330

parts of, 115–116

reducing

about, 228–230, 278–281

improper, 277–278

solving word problems that use, 281–282

subtracting

about, 274–275

like, 227–228

'front' concept, 42

G

graphs. *See* charts

greater-than (>) symbol, 35–37, 148, 159, 290–291

H

halves, 113

heavier concept, 53

hours

identifying on digital clocks, 99–101

minutes in, 99

hundreds place, 139–140

hundredths place, 307–309

I

icons, explained, 2

identifying

equivalent decimals, 313–314

equivalent fractions, 118

hours on digital clocks, 99–101

hundredths place, 307–309

matching numbers on number lines, 24

minutes on digital clocks, 99–101

missing numbers, 27, 39–40, 45, 47, 55, 59, 140

number patterns, 248–250

numbers on playing cards, 18

place values through 1,000,000, 243–245

tens and ones places, 82–85

tenths place, 305–307

improper fractions, reducing, 277–278

inches, measuring, 51–52

index cards. *See* flash cards

L

languages, second, 339

learning

address, 28–29

counting 1 to 9, 14 15

counting 10 to 20, 21–22

to measure, 50–52

numbers 1 to 40, 43–44

numbers 1 to 100, 58–60

numbers 41 to 50, 45–47

numbers 51 to 80, 53–56

phone numbers, 19–20, 24

length. *See* measuring

lesson preparation

about, 32–33

for numbers, 14–15

less-than (<) symbol, 34–35, 36–37, 148, 159, 290–291

'lighter' concept, 53

like fractions

adding, 226–227

subtracting, 227–228

lines

drawing using rules, 52

parallel, 318–319, 321

perpendicular, 318–319, 321

M

math
  mental
    about, 133
    adding 10, 145
    adding 100, 146–147
    adding in, 133–138
    comparing large numbers, 148
    counting to 1,000 by one hundreds, 138
    subtracting 10, 146
    subtracting 100, 147
    subtracting in, 133–138
  new
    adding multi-digit numbers using number lines, 256–258
    compared with old math, 8–9
    multiplying multi-digit numbers using Box method, 269–271
    multiplying two-digit numbers by single-digit numbers (Box method), 205–209
    subtracting multi-digit numbers using number lines, 263–264
    using to add/subtract large numbers, 176–179
  old
    adding multi-digit numbers through 1,000, 254–255
    compared with new math, 8–9
    multiplying multi-digit numbers, 266–269
    multiplying two-digit numbers by single-digit numbers, 200–205
    subtracting multi-digit numbers using borrowing, 260–262
  old compared with new, 8–9
mathematical expressions
  about, 285
  comparing expressions using greater-than, less-than, and equal symbols, 290–291
  completing equivalent expressions, 288–289
  with parentheses, 289–290
  solving problems
    grouped by parentheses, 287–288
    with multi-digit numbers, 286–287
    using multiple operations, 286
measuring. *See also* calculating
  angles, 320
  inches, 51–52
  learning, 50–52
  using fractions, 119
mental math
  about, 133
  adding 10, 145
  adding 100, 146–147
  adding in, 133–138
  comparing large numbers, 148
  counting to 1,000 by one hundreds, 138
  subtracting 10, 146
  subtracting 100, 147
  subtracting in, 133–138
minutes
  in an hour, 99
  identifying on digital clocks, 99–101
missing numbers, identifying, 39–40, 45, 47, 55, 59, 140
mixed numbers, 117–118, 158, 275–277, 298–301
money
  about, 149
  counting change, 150–153
  decimals and, 304
  dimes, 151–152

number lines
  adding
    large numbers using, 86–88, 176–177
    multi-digit numbers using (new math),
      256–258
    numbers through 20 using, 134–135
    using, 65–66
  checking your work using, 179–180
  comparing numbers on, 33
  counting numbers on, 18–19, 23–24, 25
  identifying tens on, 83–84
  subtracting
    large numbers using, 91–93, 177–179
    multi-digit numbers using (new math),
      263–264
    numbers through 20 using, 136–137
    using, 70–71
number patterns, identifying, 248–250
numbers
  about, 13
  adding
    with decimal points, 310
    through 20 mentally, 134–136
    using decomposition, 171–172
    zero to, 74, 75–76
  comparing
    about, 31–40
    1 through 80, 55
    1 through 100, 59–60
  counting
    on number lines, 18–19, 23–24, 25
    1 through 30, 38–40
  dividing with decimal points, 314–316
  even, 247–248
  factoring, 240–241
  identifying on playing cards, 18
  learning 1 to 40, 43–44

learning 1 to 100, 58–60
learning 41 to 50, 45–47
learning 51 to 80, 53–56
lesson preparation, 14–15
mixed, 117–118, 158, 275–277, 298–301
multi-digit, 251–272
  about, 251
  adding
    numbers through 1,000 (old math),
      254–255
  adding three rows of, 252–254
  adding through 10,000, 255–256
  adding using number lines (new math),
    256–258
  adding using rounding, 258–260
  adding with regrouping, 252
  dividing, 271–272
  multiplying (old math), 266–269
  multiplying using Box method (new
    math), 269–271
  solving problems with, 286–287
  subtracting, 260–265
  subtracting using borrowing (old math),
    260–262
  subtracting using number lines (new
    math), 263–264
  subtracting using rounding, 264–265
multiplying with decimal points,
  312–313
odd, 247–248
phone, 19–20, 24
pointing out, 16
prime, 241–242
relationship between 1 through 10 and
  11 through 20, 27
rounding, 217–219
subtracting
  with decimal points, 311

348   Teaching Your Kids New Math, K–5 For Dummies

through 20 mentally, 136–138

zero from, 75

writing, 16–17

O

objects

categorizing, 42–43

counting, 16, 17, 27, 40

weighing, 53

odd numbers, 247–248

old math

adding multi-digit numbers through 1,000, 254–255

compared with new math, 8–9

multiplying multi-digit numbers, 266–269

multiplying two-digit numbers by single digit numbers, 200–205

subtracting multi-digit numbers using borrowing, 260–262

one hundreds, counting to 1,000 by, 138

1

dividing by, 194

multiplying by, 187

prime numbers and, 241

1,000

adding multi-digit numbers through (old math), 254–255

counting to by one hundreds, 138

1,000,000, identifying place values through, 243–245

100

addition mentally, 146–147

dividing through, 209

multiplying numbers through, 204–205

rounding numbers through, 219

subtraction mentally, 147

ones place

about, 139–140

identifying, 82–85

operations

addition

of 1 minute to times with 50 minutes, 100–101

of 10 mentally, 145

of 100 mentally, 146–147

about, 237–238

creating equivalent expressions using, 160–161, 162–163

of factions, 274

flash cards through 10, 95–96

identifying tens and ones places, 82–85

identifying tens and ones places using worksheets, 84–85

identifying tens on number lines, 83–84

of large numbers using boxes, 88–89, 141–142

of large numbers using number lines, 86–88

of large numbers using open number lines, 176–177

of large numbers without regrouping, 140–144

of large numbers without using boxes, 142

of like fractions, 226–227

in mental math, 133–138

mixing with subtraction problems on same worksheet, 95

of multi-digit numbers through 1,000 (old math), 254–255

of multi-digit numbers through 10,000, 255–256

of multi-digit numbers using number lines (new math), 256–258

using tables, 185–186

by zero, 187–188

parentheses and, 238

solving problems using multiple, 286

subtraction

of 10 mentally, 146

of 100 mentally, 147

about, 237–238

addition and, 73

creating equivalent expressions using, 161–163

flash cards through 10, 95–96

of fractions, 274–275

identifying tens and ones places, 82–85

identifying tens and ones places using worksheets, 84–85

Identifying tens on number lines, 83–84

of large numbers using boxes, 93–94, 143–144

of large numbers using number lines, 91–93, 177–179

of large numbers without boxes, 94

of large numbers without regrouping, 140–144

of large numbers without using boxes, 144

of like fractions, 227–228

in mental math, 133–138

mixing with addition problems on same worksheet, 95

of multi-digit numbers using borrowing (old math), 260–262

of multi-digit numbers using number lines (new math), 263–264

of multi-digit numbers using rounding, 264–265

of numbers through 20 mentally, 136–138

of numbers through 100 without regrouping, 90, 94

of numbers with decimal points, 311

with regrouping, 165–180

through 20, 106–108

of unlike fractions, 296

using flash cards, 71–72

using index cards, 69–70

using new math for large numbers, 176–179

using number lines, 70

using straws, 68–69

without regrouping, 172

on worksheets, 73

on worksheets using number lines, 70–71

of zero from numbers, 75, 76, 77

ovals, 56–58

P

parallel lines, 318–319, 321

parentheses

expressions with, 289–290

operations and, 238

solving problems grouped by, 287–288

pennies, counting on worksheets, 150

performing, equivalent expressions, 288–289

perimeter

calculating for triangles, 324

calculating of rectangles, 321–322

perpendicular lines, 318–319, 321

phone numbers, learning, 19–20, 24

pie charts, reading data on, 224–226

place values, identifying through 1,000,000, 243–245

placing commas, in large numbers, 246

work, checking your, 175–176, 179–180,
    196–197, 216
worksheets
  adding
    numbers through 20 using, 135–136
    through 20 using, 106
    using, 64, 65, 68
    zero on, 76
  comparing numbers on, 37, 40, 48,
    56, 60
  completing
    division on, 196
    multiplication on, 189
  counting pennies on, 150
  identifying tens and ones places using,
    84–85
  subtracting
    numbers through 20 using, 138
    through 20 using, 108
  using, 70–71, 73
  zero on, 77
  timed, 96
writing
  dates, 155
  missing numbers, 27
  numbers, 16–17

Y

year, months of the, 154

Z

zero
  about, 74
  adding to numbers, 74, 75–76
  dividing by, 194
  multiplying by, 187–188
  subtraction from numbers, 75, 76, 77

Notes

Notes

About the Author

Dr. Kris Jamsa is the author of 118 books, mostly on computers and programming, but also several on pre-K through fifth-grade learning. Jamsa has two PhDs — one in computer science and another in education — and five master's degrees (business administration, project management, computer science, information security, and education). He has an undergraduate degree in computer science from the United States Air Force Academy.

Jamsa was the founder of Head of the Class, an online learning portal for pre-K through fifth-grade learners. His research interests include all things computer, as well as ways in which people can use technology to improve learning.

He and his wife Debbie live on a ranch in Prescott, Arizona, with their dogs, cats, and horses.

Dedication

To Debbie: How much I love you is something that I can't explain using math!

Author's Acknowledgements

Writing a book on math is challenging. Making a book on math easy to understand takes a great team of editors, illustrators, and designers. Putting this book together was great fun thanks to the efforts of the Wiley team. I want to thank Kathleen Kelley, this book's technical editor, for her many contributions and insights. Kathleen is one of the smartest educators I know, and her knowledge was invaluable. I want to thank Vicki Adang for teaching me the Dummies process and for patiently guiding me through this book's early chapters. In addition, I want to thank Marylouise Wiack, this book's copy editor. Not only did she fix my grammar, but she also often improved the wording of my instructions by making them easier to understand. Finally, all through this book, Charlotte Kughen, this book's development editor, kept me on track and improved every aspect. Charlotte's unselfish contributions made this a much better book. I sincerely appreciate all of their efforts and contributions.

Publisher's Acknowledgments

Acquisitions Editor: Jennifer Yee

Project Editor: Charlotte Kughen

Copy Editor: Marylouise Wiack

Technical Editor: Kathleen Kelley

Production Editor: Saikarthick Kumarasamy

Cover Image: © fizkes/Shutterstock

Take dummies with you everywhere you go!

Whether you are excited about e-books, want more from the web, must have your mobile apps, or are swept up in social media, dummies makes everything easier.

Leverage the power

Dummies is the global leader in the reference category and one of the most trusted and highly regarded brands in the world. No longer just focused on books, customers now have access to the dummies content they need in the format they want. Together we'll craft a solution that engages your customers, stands out from the competition, and helps you meet your goals.

Advertising & Sponsorships

Connect with an engaged audience on a powerful multimedia site, and position your message alongside expert how-to content. Dummies.com is a one-stop shop for free, online information and know-how curated by a team of experts.

- Targeted ads
- Video
- Email Marketing
- Microsites
- Sweepstakes sponsorship

20 MILLION PAGE VIEWS
EVERY SINGLE MONTH

15 MILLION UNIQUE
VISITORS PER MONTH

43% OF ALL VISITORS ACCESS THE SITE
VIA THEIR MOBILE DEVICES

700,000 NEWSLETT SUBSCRIPTI
TO THE INBOXES OF
300,000 UNIQUE INDIVIDUALS EVERY WEEK

# of dummies

Custom Publishing

Reach a global audience in any language by creating a solution that will differentiate you from competitors, amplify your message, and encourage customers to make a buying decision.

- Apps
- Books
- eBooks
- Video
- Audio
- Webinars

Brand Licensing & Content

Leverage the strength of the world's most popular reference brand to reach new audiences and channels of distribution.

For more information, visit dummies.com/biz

PERSONAL ENRICHMENT

Staying Sharp dummies

9781119187790
USA $26.00
CAN $31.99
UK £19.99

Facebook dummies

Carolyn Abram

9781119179030
USA $21.99
CAN $25.99
UK £16.99

Guitar dummies

Mark Phillips
Jon Chappell

9781119293354
USA $24.99
CAN $29.99
UK £17.99

Investing dummies

Eric Tyson, MBA

9781119293347
USA $22.99
CAN $27.99
UK £16.99

Beekeeping dummies

Howland Blackiston

9781119310068
USA $22.99
CAN $27.99
UK £16.99

Digital Photography dummies

Julie Adair King

9781119235606
USA $24.99
CAN $29.99
UK £17.99

Meditation dummies

Stephan Bodian

9781119251163
USA $24.99
CAN $29.99
UK £17.99

Pregnancy ALL-IN-ONE dummies

9781119235491
USA $26.99
CAN $31.99
UK £19.99

Samsung Galaxy S7 dummies

Bill Hughes

9781119279952
USA $24.99
CAN $29.99
UK £17.99

iPhone dummies

Edward C. Baig
Bob "Dr. Mac" LeVitus

9781119283133
USA $24.99
CAN $29.99
UK £17.99

Crocheting dummies

Karen Manthey
Susan Brittain

9781119287117
USA $24.99
CAN $29.99
UK £16.99

Nutrition dummies

Carol Ann Rinzler

9781119130246
USA $22.99
CAN $27.99
UK £16.99

PROFESSIONAL DEVELOPMENT

Windows 10 dummies

Andy Rathbone

9781119311041
USA $24.99
CAN $29.99
UK £17.99

AutoCAD dummies

Bill Fane

9781119255796
USA $39.99
CAN $47.99
UK £27.99

Excel 2016 dummies

Greg Harvey, PhD

9781119293439
USA $26.99
CAN $31.99
UK £19.99

QuickBooks 2017 dummies

Stephen L. Nelson, MBA, CPA, MS in Taxation

9781119281467
USA $26.99
CAN $31.99
UK £19.99

macOS Sierra dummies

Bob "Dr. Mac" LeVitus

9781119280651
USA $29.99
CAN $35.99
UK £21.99

LinkedIn dummies

Joel Elad, MBAs

9781119251132
USA $24.99
CAN $29.99
UK £17.99

Windows 10 ALL-IN-ONE dummies

Woody Leonhard

9781119310563
USA $34.00
CAN $41.99
UK £24.99

SharePoint 2016 dummies

Rosemarie Withee
Ken Withee

9781119181705
USA $29.99
CAN $35.99
UK £21.99

Fundamental Analysis dummies

Matt Krantz

9781119263593
USA $26.99
CAN $31.99
UK £19.99

Networking dummies

Doug Lowe

9781119257769
USA $29.99
CAN $35.99
UK £21.99

Office 2016 dummies

Wallace Wang

9781119293477
USA $26.99
CAN $31.99
UK £19.99

Office 365 dummies

Rosemarie Withee
Ken Withee
Jennifer Reed

9781119265313
USA $24.99
CAN $29.99
UK £17.99

Salesforce.com dummies

Liz Kao
Jon Paz

9781119239314
USA $29.99
CAN $35.99
UK £21.99

Coding dummies

Nikhil Abraham

9781119293332
USA $29.99
CAN $35.99
UK £21.99

dummies.com

dummies
A Wiley Brand